Great Natural History Books
and their Creators

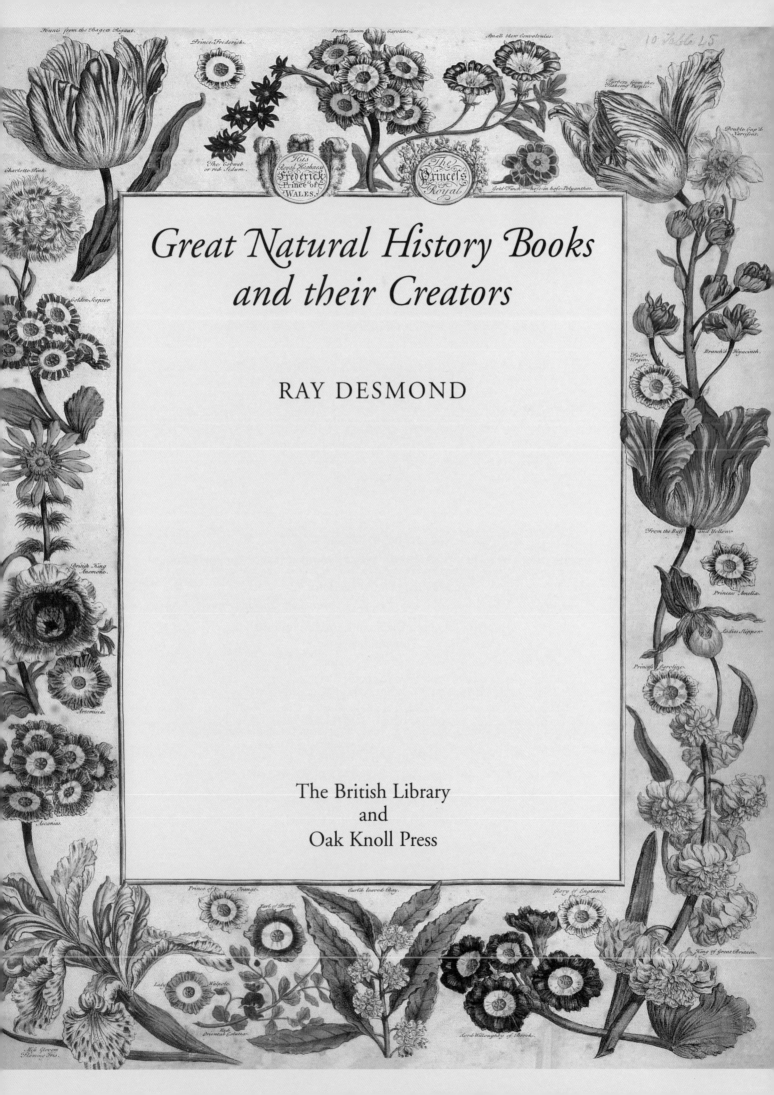

Great Natural History Books and their Creators

RAY DESMOND

The British Library
and
Oak Knoll Press

For Biddy

First published in 2003 in Great Britain by The British Library
96 Euston Road, London NW1 2DB, UK

First published in 2003 in the United States of America by Oak Knoll Press
310 Delaware Street, New Castle, DE 19720, USA

Text © 2003 Ray Desmond
Illustrations © 2003 The British Library Board and other named copyright holders

Cataloguing-in-Publication Data
A CIP record for this book is available from The British Library

Library of Congress Cataloging-in-Publication Data
Desmond, Ray.
Great natural history books and their creators / Ray Desmond.–1ˢᵗ ed.
p. cm.
Includes bibliographical references (p.).
ISBN 1-58456-090-8
Natural history literature–History. 2. Natural history–Bio-bibliography. I. Title.

QH13.45.D47 2003
508–dc21   2002030716

ISBN 0 7123 4774 7 (British Library)
ISBN 1 58456 090 8 (Oak Knoll)

Designed and typeset by Gillian Greenwood
Printed in Singapore by Craft Print

Contents

Robert John Thornton *The Temple of Flora*, London 1799–1807.
Aesculapius, Flora, Ceres and Cupid honouring the bust of Linnaeus. Painted by John Opie and John Russell.
Engraved by Caldwell, 1806. 'The introduction of Flora, Ceres and Aesculapius, is emblematic of the advantage
derived from a study of the science of Botany, as in the works of Linnaeus … Cupid is represented in allusion
to the sexual system, invented by Linnaeus' (Thornton). Linnaeus was a dominant influence in
botanical literature during much of the eighteenth century. B.L.10.Tab.40.

Preface

When we enjoy a book or admire its illustrations we seldom give a thought to any problems that might have been encountered during its production. Why should we? Most books have a relatively easy passage from the author's desk to the booksellers' shelves. Some, however, endured a painful or protracted birth. To know, for instance, that T. E. Lawrence had to rewrite the greater part of *The seven pillars of wisdom* when his manuscript was lost during a train journey, or that Thomas Carlyle faced the same daunting task when a maidservant accidentally used his draft of the first volume of *The French revolution* to light a fire, makes us regard the unfortunate author and his book with a greater respect.

With this thought in mind, I have selected a number of books in the natural history field with an interesting or distressing story to tell from the moment of their conception to their publication. Some emerged during the seventeenth and eighteenth centuries from voyages of discovery or the labours of colonial officials and solitary travellers. Most of my examples have been chosen from those botanical works which I know best. The publication of splendid flower books, now collectors' items, often originated from the introduction of new plants from abroad. Botanic gardens were created to study this exotic flora, gardeners tested their skills in cultivating the plants, and botanical artists drew them for proud owners.

What all these books have in common is commitment on the part of authors and artists, a determination to see their work in print despite the hazards of travel, lack of support or financial problems. Overreaching ambition reduced some authors to penury. Some books appeared posthumously.

Wherever desirable I have tried to set the books within the context of contemporary literature. So, for example, in the chapter on the Revd Gilbert White and the Revd Keble Martin I also mention the writings of other clerics. In the chapter on nature printing I note earlier examples of this genre before discussing Moore's and Lindley's *Ferns of Great Britain and Ireland*. An introductory chapter surveys the methods (and frustrations) of publishing and selling natural history books up to Victorian times.

My task has been made easier by the scholarly researches of authors whose books and articles are acknowledged in the Bibliography. I must thank Charles Hatt for introducing me to the extraordinary story of J. B. A. Beringer, and also Ron Cleevely for suggesting some useful references about him. I am grateful to John Flanagan and his helpful staff at the library of the Royal Botanic Gardens, Kew, to Gina Douglas, librarian of the Linnean Society, and to the staff of the Botany library at the Natural History Museum, London. The majority of the books I have chosen for illustration come from the stock of The British Library, unless otherwise stated.

1 The book trade

Many booksellers were established in London by the mid-seventeenth century, most of them located in the neighbourhood of St Paul's Cathedral. Some published as well as sold books. Publishing had once been a role of printers, such as Richard Banckes, John Byddell, Garet Dewes, William Powell, Thomas Purfoot, Peter Traveris and Robert Wyer, who during the sixteenth century produced some well-known herbals. Towards the end of the century the function of bookselling/publishing gradually separated from that of printing.

The book trade declined during the Civil War and suffered a serious setback in the Great Fire of 1666, which destroyed many bookshops and their stocks. With a slow recovery booksellers were naturally reluctant to undertake the publication of costly books on plants and animals. The newly-formed Royal Society, which had several distinguished naturalists as Fellows, might have been expected to promote or support their publication. However, during the 1680s it had little money at its disposal and its administration was ineffectual until Hans Sloane became Secretary in 1693. The experience of John Ray (1627–1705), probably the Society's most eminent naturalist, is typical of the frustrations endured by scientific authors.

Ray's greatest contribution to taxonomic botany was his three-volume survey of the world's flora, *Historia plantarum* (1686–1704), arranged according to his own classification. He regretted that no illustrations complemented its 3000 pages of text. He himself could not afford to pay for engravings, nor would his bookseller agree to do so. 'The times indeed of late have not been very propitious to the bookseller's trade.'[1] He received £30 and 20 copies for each of the three volumes, perhaps not an unreasonable fee for a work which, it is believed, attracted fewer than 200 subscribers.

Its disappointing sales may account for the lack of enthusiasm on the part of his bookseller, Samuel Smith, to bring out in 1698 a new edition of Ray's *Methodus plantarum nova* (1682). When Petrus Hotton, a Dutch botanist, heard of this protracted delay in 1700, he offered to have the book printed in Holland. Of course Ray agreed, and by 1703 the Amsterdam publisher had printed 1100 copies of *Methodus plantarum emendata et aucta*. It was sold in Britain under the imprint of Smith and Walford, who also handled Ray's *Synopsis methodica avium et piscium*. It was only after Smith's death, when his business was acquired by William Innys, that the *Synopsis* eventually appeared, in 1713, by which time Ray had been dead for eight years.

Authors resented the unsatisfactory treatment they got from booksellers, who in turn complained about the piracy of books in which they had invested money. With the

---

1. Ray to E. Lhwyd, 2 August 1689. R.W.T. Gunther *Further correspondence of John Ray*, 1928, p.191.

expiry of the Licensing Act in 1695 neither authors nor publishers had any protection in law. Few new books were being registered with the Stationers' Company, whose bye-laws were often ignored. Booksellers petitioned Parliament in 1703, 1706 and again in 1709 for a new licensing Act to curb the practice of the unauthorized printing and publishing of their more popular books and pamphlets. Parliament responded in 1709 with a Copyright Act, which came into force in April 1710. Under this Act authors of books already in print were to enjoy copyright for 21 years, but legal protection for new books was restricted to 14 years, with an extension of another 14 years if authors were still alive at the end of that period. This was not a provision that booksellers welcomed, since they regarded new publications as their property and persuaded many authors to sell their copyright to them. Those who refused to comply usually received a cursory service from the trade in advertising and selling their works.

The Act failed to curb the pirates, who were prepared to face possible lawsuits rather than forgo exploiting a profitable demand for cheap books. Piracy was an extension of the acceptable practice of plagiarism. E. Sweerts's *Florilegium* (1612) copied from de Bry's *Florilegium* (1611), which in turn had appropriated plates from P. Vallet's *Le jardin du Roy très Chrestien Henry IV* (1608). The books of de Bry and Nicolas Robert were a quarry for M. S. Merian's *Neues Blumen Buch* (1680). She herself received no acknowledgement in E. Albin's *Natural history of English insects* (1720), which copied her drawings of insects. Albin's book was anonymously pirated in 1754, 1776 and 1791. James Petiver had no scruples about using engravings in Rumphius's two books on the natural history of Amboina. Rogue printers and publishers were particularly active during the 1730s. The popularity of Robert Furber's *Twelve months of flowers* (1732) spawned an unauthorized imitation. The Blackwells took legal action to block the appearance of a pirated version of *A curious herbal* (1737–39). Philip Miller complained that his *Gardeners kalendar* (1732) had been 'pyratically printed' in 'a spurious edition' of his *Gardeners dictionary*. Booksellers were powerless to prevent Continental firms printing their books without permission. In 1781 G. F. Cassell complained to Sir Joseph Banks about this practice by German printers who were protected by 'some petty German prince or other'.

The ill feeling between authors and booksellers persisted throughout the eighteenth century. M. S. Merian told J. Petiver that she was reluctant to employ a bookseller to distribute her books because of the 50 per cent commission he would demand. Horace Walpole voiced a similar complaint: 'if I do not allow them [i.e. booksellers] ridiculous profits, [they] will do nothing to promote its sale [i.e. one of his Strawberry Hill books].' Richard Bradley, who wrote prodigiously not only to earn a living but also to pay out-standing debts, also suffered from the hard deals struck by booksellers. 'I want very much to gett out of the booksellers hands,' he confessed to Sir Hans Sloane.

Both Richard Bradley and his successor as Professor of Botany at Cambridge, John Martyn, neglected their official university responsibilities. Professor Humphrey Sibthorp was equally ineffectual at Oxford. The current enthusiasm for natural history was generated not by the universities but by amateurs from all strata of society – tradesmen, clergymen, doctors, and wealthy patrons like Lord Bute, the Duchess of Portland and Sir Joseph Banks. There now appeared to be a market for well-illustrated books, which authors like Eleazer Albin and George Edwards were eager to exploit with their bird books. It was, however, a risky business, as John Martyn found out by the poor response to his attractively illustrated *Historia plantarum rariorum* (1728–37). Thomas Knowlton probably had this work and the *Hortus Elthamensis* (1732) in mind when he expressed his concern to Richard Richardson in 1736 about books 'made for pompe to fill a library & more for outwarde show than reale use'. J.J. Dillenius, the author of *Hortus Elthamensis*, almost certainly concurred. He told Richardson that James Sherard, whose garden plants he described in two folio volumes, had insisted on this large format and additional plates 'to make it look bigger and more pompous'.

When Sherard reneged on his promise to underwrite the cost of paper and the plates for the *Hortus Elthamensis*, Dillenius himself engraved all the flower drawings he had made and still lost over £200 on the venture. Production costs could be considerably reduced by dispensing with the services of a professional engraver. For that reason Mark Catesby, George Edwards, Eleazar Albin, Elizabeth Blackwell, James Bolton and other authors learned the craft of engraving.

But such economies did not always ensure eventual publication. As a prestigious name could help sell a book, patrons were sought for support and, hopefully, for donations. A grateful James Bolton dedicated his *History of fungusses* (1788–91) to his patroness, the Duchess of Portland. With financial assistance from several patrons, Albin had been able to publish a *Natural history of birds* (1731–38). Wealthy patrons not only gave money but also generously allowed access to their important collections. The Duchess of Portland had a botanical garden, a menagerie and a fine library at Bulstrode Park. Sir Hans Sloane opened his collections (destined to be the nucleus of the future British Museum) to all bona-fide researchers, Mark Catesby and Elizabeth Blackwell among them.

It was customary to dedicate books to eminent people both as an expression of gratitude and as a means of stimulating sales. As well-known figures in the natural history world, Sir Hans Sloane, Richard Mead and Sir Joseph Banks were often extolled in dedications. Royal approval was assumed in any dedication to a reigning monarch or his spouse, or to the heir apparent. Mark Catesby dedicated the first volume of his *Natural history of Carolina* (1729) to Queen Caroline and the second to Augusta, Princess of Wales. Sir Hans Sloane respectfully offered volume one of *A voyage to the islands Madera, Barbados…* (1707–25) to Queen Anne and volume two to her successor, George I. When an English edition of her book on the insects of Surinam seemed possible, M.S. Merian wanted to dedicate it to Queen Anne. All royal dedications were suitably deferential, but Robert Thornton reached the ultimate in flattery in his dedication of *The Temple of Flora* to Queen Charlotte. After that of royalty, the recommendation of peers was sought. John Hill's *British herbal* (1756) and Philip Miller's eighth edition of his *Gardeners dictionary* (1768) acknowledged the encouragement of the Duke

An advertisement in *The present state of Europe: or, The historical and political monthly mercury* for May 1698 announcing a new book by William Salmon. Such proposals were usually published as a separate leaflet for distribution. His *Botanologia: the English herbal* did not appear until 1710. Bodleian Library, Oxford.

of Northumberland. Richard Weston's *English flora* (1775) and the first volume of William Curtis's *Flora Londinensis* (1777) singled out Lord Bute for a graceful tribute. The fulsome praise of these dedications call to mind Samuel Johnson's waspish definition of a dedicator: 'one who inscribes his work to a patron with compliments and servility'.

The inclusion of the names of distinguished dedicatees was calculated to impress potential book purchasers, among them some notable bibliophiles such as Sir Hans Sloane and Sir Joseph Banks. Sloane amassed about 50,000 books, manuscripts and volumes and portfolios of drawings, with natural history subjects predominating. After the death in 1754 of Richard Mead, whose collections included drawings by M.S. Merian, the sale of his library lasted 28 days. The disposal of the natural history items in Lord Bute's library in 1794 took place over ten days. Sir Joseph Banks's library was supervised by Jonas Dryander, who compiled a catalogue of it which was published in five volumes between 1798 and 1800. Banks had 250 copies printed for distribution to friends and scholars. The formation of these great libraries during the Age of Enlightenment benefited both authors and booksellers.

Patrons played an important role in promoting books, but publicity, then as now, was a key factor. Authors or booksellers usually distributed prospectuses or proposals giving brief information on a book's contents, price and availability. Newspapers such as the *Public Advertiser* and the *London Chronicle* regularly announced new and forthcoming books. Advertisements were slipped into appropriate books (James Petiver's *Herbarii Britannici* was announced at the end of John Ray's *Synopsis methodica* (1713)), but the prospectus remained the preferred medium of publicity for many years. News of Audubon's *Birds of America* and Sibthorp's *Flora Graeca* was promulgated in that way even as late as the nineteenth century.

The prospectus was a way of sounding out the market by inviting subscriptions before publication. If no bookseller was willing to risk publishing a book, then an impecunious author had to rely on subscribers to share the cost of publication. To entice them his prospectus might offer favourable rates and protracted payment. As a general rule subscribers made an initial deposit, with the balance being paid on completion of the book. Elizabeth Blackwell accepted payment for *A curious herbal* (1737–39) in three instalments. Robert Plot's *Natural history of Staffordshire* (1686) traded on the attractions of exclusivity by limiting its availability to subscribers only. Plot also promised that 'all subscribers whatever, shall have their names registered to Posterity, in a Printed Catalogue annex'd to the Book, as Benefactors to the work, and Promoters of the Honour of their Country'.

Human vanity was further gratified by the dedication of individual plates in a book to those people prepared to pay for this honour. The Royal Society charged its Fellows a guinea for their names to be engraved on plates in John Ray's *History of fishes* (1686). James Petiver expected two guineas for a crudely printed dedication label pasted to a plate in his *Catalogue of Mr Ray's English herbal* (1713). It comes as no surprise that he failed to find subscribers for every plate. The extra money raised by this means did not keep Robert Morison and John Martyn solvent. The former claimed he had been 'ruinated' by his *Plantarum historiae universalis Oxoniensis* (1680–99) and the latter's *Historia plantarum rariorum* (1728–37) ceased publication after only five parts and 50 plates.

Subscribers were sometimes found by word of mouth, usually by obliging friends of the author. Thomas Knowlton, for instance, was indefatigable in getting subscribers for Mark Catesby. Audubon exhibited his original bird paintings at different venues in Britain to seduce possible purchasers of *Birds of America*. The subscription lists that were often included in books are very revealing about the social status of the subscribers of natural history works. In the eighteenth century they came, not unexpectedly, from the nobility and landed gentry and also included academics, doctors and apothecaries, members of the clergy, a few gardeners and some booksellers. The number of subscribers fluctuated, for example, from 25 for J. Sibthorp's costly *Flora Graeca* (1806–40), 79 for

Robert Furber *Twelve months of flowers*, 1730–32. Lists of subscribers were frequently added to books to flatter them and as commendation for the books. This example is exceptionally elegant with a floral border designed by Peter Casteels and engraved by Henry Fletcher. B.L.10.Tab.45.

OPPOSITE
Moses Harris *The Aurelian*, 1766.
Most of its 44 plates were dedicated to a patron or subscriber, this particular one to the Duchess of Richmond. Harris very astutely dedicated the first plate to Lord Macclesfield, President of the Royal Society, 'and to the rest of the Fellows', some of whom were likely purchasers. B.L.459.f.11.

Mrs P. Bury's *A selection of Hexandrian plants* (1831–34), 318 for W. Curtis's *Flora Londinensis* (1777–98), 407 for P. Miller's *Gardeners dictionary* (1731) and 435 for R. Furber's *Twelve months of flowers* (1730–32) to 742 for A. Hunter's edition of J. Evelyn's *Sylva* (1776).

The number of subscribers was obviously crucial to a book's survival. Sir James E. Smith's *Exotic botany* (1804–05) stopped after two volumes 'for want of public patronage'. J.C. Loudon was compelled to discontinue his *Illustrations of landscape gardening and garden architecture* (1830–33) after the third part due to insufficient subscriptions, despite the fact that he offered it to the trade and gardening fraternity at a discount. More than half of Audubon's English subscribers failed to buy the entire work. On that occasion an economic depression in the mid-1830s was partly responsible, but usually it was protracted and erratic publication that caused many subscribers to lose interest.

To Her Grace the Dutchess of Richmond.

This Plate is humbly Dedicated by her Graces most Obliged & Obed.t Hum.ble Serv.t

Moses Harris

During the eighteenth century large and generously illustrated books frequently appeared in weekly, monthly or irregular parts, a practice welcomed by subscribers, who were thus able to pay in instalments. Moreover the publisher could spread his costs over a longer period and allow himself more time to prepare any copper plates. Sir Joseph Banks advised the East India Company to adopt this method in publishing *Plants of the coast of Coromandel* (1795–1820). In this way, he said, the success of the book would be 'tried at a moderate expense, which would, in the case of failure [were] the whole to be published together, involve a considerable sum of money'.

Each part-issue contained some text and a few plates stitched together within a paper wrapper. Subscribers were responsible for the ultimate binding of these parts with an opportunity to rearrange the plates, as recommended, for instance, by W. Curtis in respect of his *Flora Londinensis* (1775–98). Part-issues were an innovation enthusiastically endorsed by the *Grub Street Journal* for 11 October 1732: 'This method of weekly publication allows multitudes to peruse books, into which they would otherwise never have looked.'

Part-issues were tentatively introduced during the closing decades of the seventeenth century, becoming popular from the 1730s onwards, when they made huge profits for publishers. *The compleat herbal or, the botanical institutions of Mr Tournefort*, promoted by a consortium of London booksellers, was one of the first botanical books in this country to be published in parts. Its preface justified this then still novel mode of publication, referring to 'the expence … which otherwise would have amounted to too considerable a sum', and pointing out that 'if the book does not answer the expectations, nor hit the taste of the curious' it could easily be terminated. Forty-one one shilling parts, issued between 1716 and 1730, cumulated into two quarto volumes. Issues of Mrs Blackwell's *A curious herbal* (1737–39) came out regularly for 125 consecutive weeks. J. Sowerby's and Sir James E. Smith's *English botany* (1790–1814) reached a record 267 parts with another 83 in its supplement (1829–66). The prize, however, has to go to the *Flora Danica*, which struggled on, albeit intermittently, for 122 years. An inherent drawback to this otherwise popular way of publication was the uncertainty of its regular appearance and its eventual completion.

'Natural history of all sorts is much in demand,' wrote William Sherard in November 1720 to Richard Richardson. Its popularity had not waned when Peter Collinson told Linnaeus in April 1747 that 'we are very fond of all branches of natural history. They sell the best of any books in England.' Natural history was still predominantly the recreation and study of the leisured classes and academics, but gradually its appeal spread; in rural parishes the clergy were attentive observers of the countryside. It attracted a new breed of knowledgeable gardeners like Philip Miller and Thomas Knowlton. Europe was enthralled by new species of fauna and flora discovered in its colonies. Peter Collinson, James Sherard, John Fothergill, J.C. Lettsom and other gardening enthusiasts coveted newly introduced exotics. This cornucopia of new plants needed a simple system for identifying and classifying them. When the Swedish naturalist Linnaeus devised such a scheme, it led to a stocktaking of the natural world through assiduous collecting, labelling and listing.

This fashionable trend in taxonomic studies was reflected in an increase of natural history works. R.B. Freeman's sampling of those published between 1495 and 1800 (*British natural history books* (1980)) shows that their publication accelerated during the eighteenth century. After a slow start, with 12 titles in the sixteenth century, expanding to 97 in the seventeenth century, the eighteenth century begins with 27 for the first two decades, 15 in the 1720s, 23 in the 1730s, 24 in the 1740s, 32 in the 1750s, 28 in the 1760s, 57 in the 1770s, 66 in the 1780s and 104 in the 1790s. The obsessive writer John Hill dominated the 1750s with 11 works; his *Vegetable system* (1759–75), a massive undertaking in 26 folio volumes, beggared him and left his widow impoverished. In order to keep his *Gardeners dictionary* affordable for gardeners, Philip Miller published its illustrations separately. The 1750s were distinguished by Miller's dynamic development of the Chelsea Physic Garden and the establishment of a botanical garden at Kew. During this decade the *Annual Register* and the *Gentleman's Magazine* started to give space to natural history topics. The *Critical Review* in 1763 affirmed that 'natural history is now by a kind of national establishment, become the favourite study of the times'. 'Anything in the naturalist way now sells well,' Gilbert White assured his brother John in March 1775.

Booksellers prospered. They purchased authors' rights, subcontracted printing, platemaking and binding, and marketed the books. A group of them got together to advertise new publications. One of London's leading booksellers/publishers was Benjamin White, Gilbert White's younger brother, with premises in Fleet Street. His customers included the naturalists Thomas Pennant, Thomas Martyn, Richard Pulteney and the great bibliophile Earl Spencer. He reissued George Edwards's edition of Catesby's *Natural history of Carolina* (1771) and published John Lightfoot's *Flora Scotica* (1777) and works by W. Curtis, T. Martyn and J.E. Smith. Sir Joseph Banks patronized George Nicol, a bookseller of Pall Mall, who published W. Aiton's *Hortus Kewensis* (1789), F. Bauer's *Delineations of exotick plants ... at Kew* (1796–1803) and W. Roxburgh's *Plants of the coast of Coromandel* (1795–1820).

The international trade in books expanded during the eighteenth century. In 1729 Thomas Knowlton informed a friend that the first part of J.C. Buxbaum's *Plantarum minus cognitorum*, published in St Petersburg, was now on sale in England. Until his death in 1780 John Nourse was London's leading importer of foreign literature, occasionally purchasing multiple copies of individual titles. A great deal of business was done with Holland, with booksellers visiting each other's country. Germany was another important market. At Nuremberg C. J. Trew (1675–1769), a physician and a collector of books, paintings and natural history specimens, employed flower painters to illustrate his *Hortus nitidissimus* (1750–86) and *Plantae selectae* (1750–73). He also published a German version of Mrs Blackwell's *A curious herbal* (1757–73). His influence in Germany as a patron of natural history and associated literature matched that of Sir Hans Sloane and Sir Joseph Banks in England. The botanist Nikolaus Joseph von Jacquin (1727–1817),

Director of the royal gardens at Schönbrunn who exercised an influence in Vienna similar to Trew's in Nuremberg, had some difficulty in selling his superbly illustrated books. He begged Sir Joseph Banks in October 1774 to find English buyers for his *Hortus botanicus Vindobonensis* (1770–76) since his booksellers were not interested. The London bookseller P. Elmsley ordered 30 copies of the first volume but only 20 of the second. The boom in the sale of expensive natural history books showed signs of decline by the close of the century. Sir Joseph Banks blamed it on the Napoleonic wars. In 1798 the Marchioness of Rockingham hesitated to pay 30 guineas 'in such sad times as these…for two volumes of botanical flowers' by Jacquin.[2]

Expensive materials (paper in particular) and higher labour costs – a London compositor could earn as much as 33 shillings a week in 1801 – inflated book prices. A committee convened by the House of Commons in 1818 was informed by a printer that 'books are a luxury, and the purchase of them has been confined to fewer people'. Illustrations absorbed a significant portion of the cost of most books. In 1813 T.N. Longman told the Committee on Acts respecting Copyrights of Printed Books that a book with engravings would cost 50 per cent more in England than in Europe. The number of copies printed was also a factor. Until the introduction of powered printing machinery, editions of more than 2000 copies were uneconomic. Some botanical works, usually those with few or no illustrations, were put out in comparatively large editions, e.g. Sir James E. Smith's *Introduction to botany* (1808) – 1500 copies; W.T. Aiton's *Hortus Kewensis* (1810–13) – 1250 copies. Books with expensive illustrations appeared in small editions, e.g. Mrs P. Bury's *Hexandrian plants* (1831–34) – 80 copies; E. Rudge's *Plantarum Guianae rariorum* (1805–06) – 150 copies; W. Roscoe's *Monandrian plants* (1824–29) – 150 copies.

In 1815 Goethe expressed astonishment at 'the mad luxury carried out in England with regard to books', quoting as an example 'a botanical work treating of pine trees' costing 80 guineas. This was A.B. Lambert's *A description of the genus Pinus* (1803–24), of which 200 copies were printed. Lambert told Sir James E. Smith in 1803 that he anticipated that he would be about £400 out of pocket. Ferdinand Bauer was so disappointed by the lack of interest in his *Illustrationes florae Novae Hollandiae* (1813–16) that he abandoned it. An enthusiastic review of it in the *Monthly Magazine* for 1813 lamented that 'the circumstances of the time, it must be allowed, are not very favourable to the prosecution of works of this nature'.

Early in the nineteenth century publishers *per se* gradually usurped the role traditionally performed by booksellers. Books on botany, zoology and related subjects featured in their catalogues, and firms like Lovell Reeve and John van Voorst were to make natural history their speciality. The demand for fine flower books dwindled, with just an occasional dazzling outburst. Orchidomania prompted John Lindley to write several attractive accounts of the genera, but according to the bookseller H.G. Bohn 'the most splendid botanical work of the present age' was James Bateman's *Orchidaceae of Mexico*

2. 12 August 1798. Letters of Sir J.E. Smith, vol.15, f.239. Linnean Society.

James Bateman *Orchidaceae of Mexico and Guatemala,* 1837–43. George Cruikshank's
wood-engraved vignette is a wry comment on the size of this elephant folio. B.L.Tab.1248.d.

and Guatemala (1837–43). Only 125 copies were printed in eight parts, selling at sixteen
guineas a set. Subscribers guaranteed the sale of 112 copies. Its 40 lithographed plates
alone cost Bateman £8000. This magnificent elephant folio is botany's riposte to
Audubon's *Birds of America.* It was now the turn of zoology to excite attention. 'You could
not pass a bookseller's from the extreme West End to Wapping without seeing new books
on zoology in every window' (Audubon, 1838). The tradition of opulent colour plate
books was now being perpetuated by bird books. Edward Donovan's *Natural history of
British birds* (1794–1819), George Graves's *British ornithology* (1811–21), P.J. Selby's
Illustrations of British ornithology (1821–34) and Sir William Jardine's *Illustrations of
ornithology* (1826–43) catered for this latest fashion. From his London residence in
Charlotte Street John Gould directed a team of artists, lithographers and colourists to
design books on the birds of Europe, Australia, America and Asia. Audubon said that
illustrations of 'insects, reptiles and fishes' enjoyed a brief vogue as well.

It was the fate of many of these de luxe volumes to be remaindered. The American
botanist Asa Gray, visiting London in 1839, jubilantly recorded in his journal that he
had 'bought a coloured copy of Wallich's *Plantae Asiaticae rariores*, 3 vols, very fine, for
£15, the publishing price was £36'. Lackington, Tegg and Bohn, three astute booksellers,
pioneered the practice of underpricing books.

James Lackington (1746–1815), formerly a journeyman shoemaker in Bristol, had
dabbled in second-hand books before opening a bookshop in 1774 in London, a city he
described as the 'grand emporium of Great Britain for books'. He was appalled to discover that booksellers who purchased publishers' unsold stock destroyed much of it in
order to maintain the full publication price of what they kept. Despite vigorous oppos-
ition from the trade, Lackington sold such surplus books at a quarter to a half of their
original price. In 1793 he moved to a large building on the corner of Finsbury Square.
A flag above the entrance announced the 'cheapest bookshop in the world'. This 'Temple
of the Muses', set out in a series of galleries on several floors, displayed the most expensive
books on the ground floor, gradually progressing to the cheapest at the top of the building.
Lackington retired, a very rich man, in his early fifties.

Thomas Tegg (1776–1845) survived several bankruptcies before making any profit out of selling remaindered books. He described himself as 'the broom that swept the booksellers' warehouses'. Later he concentrated on publishing condensed versions of popular works and reprinting in cheap editions books recently out of copyright.

Tegg admitted that Bohn surpassed him in taking advantage of the remaindered books market. Henry George Bohn (1796–1884), the son of a London bookseller, founded his own business in 1831. In 1841 he issued a mammoth *Guinea catalogue* offering over 23,000 items for sale. When he published his *Catalogue of books* of antiquarian and scientific interest six years later, he claimed to have more than half a million books in stock. Among its bargains were Horace Walpole's copy of Catesby's *Natural history of Carolina*, bound in vellum (eight guineas); Besler's *Hortus Eystettensis* (three guineas); Thornton's *The Temple of Flora*, published at £42 (six guineas); Trew's edition of Blackwell's *A curious herbal* (seven guineas); Roxburgh's *Plants of the coast of Coromandel*, published at £63 (eighteen guineas in good boards, £24 bound); Gould's *Birds of Europe*, published at £76 (£55); Audubon's *Birds of America*, published at £200 (£100 in parts, £130 bound).

Bohn never missed any promising book auction. He purchased several choice items and collections of drawings at the sale of A.B. Lambert's library in April 1842: Lambert's own copy of his *Genus Pinus*, Sibthorp's *Flora Graeca*, Jacquin's *Selectorum stirpium Americanum*, Reede tot Drakenstein's *Hortus Indicus Malabaricus*. When W. Swainson emigrated to New Zealand, Bohn lost no time in buying up the remaining copies of his *Zoological illustrations* before he left.

As well as buying surplus stock he also bought the copyright of books in order to reissue them in cheap editions. With this in mind he bought copper plates, lithographic stones and wood blocks as well as sheets of letterpress and loose plates. He reissued the revised edition of Curtis's *Flora Londinensis*. He purchased the engraved plates and spare letterpress of Sibthorp's *Flora Graeca* to print another 40 copies. When he retired in 1864 it took 40 days to sell his stock of second-hand books. In 1891 Puttick and Simpson sold by auction 'An extensive collection of attractive old engraved copper plates and a quantity of old wooden blocks collected with a view to republication by the late H.G. Bohn'.

Bohn reissued all 40 volumes of Sir William Jardine's *The naturalist's library* (1833–43). Jardine had intended them for readers who could not afford expensive books. William MacGillivray had recommended their small format to Audubon, urging him to abandon his concept of 'imperial size and regal price'. Steam printing presses, machine-made paper and stereotype plates reducing costs ushered in the era of cheap literature. Richard Groombridge was one of the first publishers to bring out really inexpensive botanical books, for example Deakin and Marnock's *Florigraphia Britannica* (1837–48) in sixpenny parts. The economist G. Porter noted in 1852 that 'books can no longer be considered, as they were in bygone times, a luxury, to be provided for the opulent, who alone as a class, would read and enjoy them'.

Technical developments in book production during the nineteenth century eventually brought about the demise of hand-colouring. Woodcuts in early herbals were sometimes coloured, although purchasers were normally expected to do their own colouring. Crispin de Passe's *A garden of flowers* (1615) instructed the reader in 'the perfect true manner of colouringe the same with their natural coloures'. Before too long, manuals guided the novice: *A book of drawing, limning, washing, or colouring of maps and prints* (1660) and *The art of painting in oyl ... to which is added the whole art and mystery of colouring maps and other prints with water coloures* (1723). Colouring was by no means mandatory. The density of parallel and cross-hatched lines in the black and white prints reproduced from the eighteenth century copper plates in *Captain Cook's florilegium* (1973), expertly rendering contours, textures and tonal values, suggests that maybe Sir Joseph Banks had never intended them to be coloured. The botanist William Sole, believing that 'good plates are injured by colouring ... endeavoured to procure such plates as need no colouring' (Preface to his *Menthae Britannicae* (1798)). The general public, however, probably subscribed to Horace Walpole's dictum that a 'want of colouring is the capital deficiency of prints'.

Applying colour to a print was not as straightforward as it might appear. First of all it had to be sized (isinglass was recommended) to reduce the absorbency of the paper. Since opaque colours obliterated finely engraved lines, transparent washes were preferred. A final coating of gum arabic or albumen gave depth to certain colours. In the preface to his *Natural history of Carolina* Mark Catesby extolled those colours 'most resembling Nature, that were durable and would retain their lustre, rejecting others very specious and shining, but of an unnatural colour and fading quality'. In his day artists had a restricted range of watercolours at their disposal. Until the mid-eighteenth century, when watercolours became available in small cakes, artists ground their own pigments from raw materials. Prussian blue was not available until 1704, cobalt until 1802 and French ultramarine until about 1815. Winsor and Newton introduced 'Chinese white' in 1834. These are facts to be borne in mind when any assessment is made of colour fidelity in old natural history books.

It was customary for publishers to employ women and youngsters as colourists. Children from an orphanage coloured the woodcuts in Salomon Schinz's *Anleitung zu der Pflanzenkenntniss* (1774). Under supervision, children coloured the aquatints in H. Repton's *Observations on the theory and practice of landscape gardening* (1803). Colourists often worked as a team, having before them a pattern print to copy. The firm of Lovell Reeve, one of the few publishers still employing colourists in the 1920s, described the routine:

> *A regular colourer prefers to work in hundreds of the same plate, one colour at a time and one plate after another, in a purely mechanical way. An expert colourer working full time may perhaps colour 200 8vo plates a week, more or less, according to the number of colours and the amount of surface and detail: but their rate varies very much with their personality.*[3]

3. 11 January 1922. Lovell Reeve Publisher's notes on the production of *Curtis's Botanical Magazine*. Royal Botanic Gardens, Kew.

Rudolph Ackermann published so much colour work at his Repository of Arts in the Strand that he maintained his own staff of colourists. Some 30 people were at one time engaged in colouring *Curtis's Botanical Magazine*. Audubon's *Birds of America* required about 50 people to colour its enormous aquatints. James Sowerby relied on his family to help colour the many parts of *English botany* (1790–1814). The unfortunate daughters and daughter-in-law of William Baxter coloured 600 sets of the 509 plates which made up his *British phaenogamus botany* (1834–43). It was not unusual for the artist to do the colouring of a small imprint. Francis Bauer exquisitely coloured all the copies of his *Delineations of exotick plants ... at Kew* (1796–1803), and Mary Lawrance applied transparent and opaque colours to her *Collection of roses from nature* (1799).

Repetitive work and low wages did not encourage colourists either to aim at or to maintain high standards. One is not surprised that the publisher George Nicol had difficulty in 1816 in recruiting any with tolerable skills. In 1762 Denmark had tried to resolve a shortage of colourists by establishing a school to train them. The firm of Benjamin Fawcett (1808–93), a colour-block maker and printer who illustrated many of Richard Groombridge's natural history books, was one of the last to run a separate colour department. Fawcett took on local girls at Driffield in Yorkshire from the age of fourteen or fifteen and trained them in stages progressing through the application of simple washes, shading and precise detail.

The quantity of hand-colouring seldom satisfied all subscribers. With any team of colourists variation in competence could be expected. When they were under pressure to meet deadlines, as with, for example, the monthly issues of *Curtis's Botanical Magazine*, fluctuations in care and accuracy were to be expected. When Sir Joseph Hooker edited it he constantly complained to the proprietor, who on one occasion tartly reminded him that

> *print-colourers are not artists & cannot do artists' work nor will the price permit artists' pay. All we can hope or profess to do is to give such a portrait of the plant as will give a fairly accurate idea of its structure & habit.*[4]

What is particularly disturbing is inconsistency in the colouring of the same plant in different copies of the same book. Nicolas Barker's meticulous examination of several copies of Besler's *Hortus Eystettensis* (1613) revealed such lapses. These imperfections, however, do not detract from the impressive achievements of the colourists, for the most part anonymous and paid a pittance.

One way of economising on the cost of hand-colouring was not to colour the entire plate. In some parts of *Curtis's Botanical Magazine* and *English botany* colour was confined to a flower and a leaf. As a rule the cost of a coloured copy of a book was twice that of a plain one. For example, one part of P. Miller's *Figures of the most beautiful ... plants* (1755–60) containing six plates sold at two shillings and sixpence plain and five shillings

---

4. F. L. Soper to Hooker, 8 April 1905. English letters, 1901–05, ff.1458–59. Royal Botanic Gardens, Kew.

coloured. The same rate was charged for a part of W. Curtis's *Flora Londinensis* (1775–98), also with six plates, but a few with superior colouring were priced at seven shillings and sixpence. An uncoloured copy of J. Lindley's *Digitalium monographia* (1821) could be bought for four guineas, a coloured one cost two guineas more. A coloured copy of B. Besler's *Hortus Eystettensis* (1613) cost ten times that of a plain copy, the justification being the exceptionally high standard attained by the colourist.

Hand-colouring could not compete economically with Victorian chromo-lithography and colour wood blocks. *Curtis's Botanical Magazine* remained loyal to it until 1948, when a dearth of colourists forced the periodical to adopt mechanical colour printing. It is occasionally revived to add distinction to limited editions and presentation copies. Forty copies of supplements eight and nine, in 1960 and 1962, of H.J. Elwes's *Monograph of the genus Lilium* were hand-coloured by the botanical artist Margaret Stones.

The Temple of the Muses, on the corner of Finsbury Square, James Lackington's highly successful bookshop. On its opening in 1793 a mail coach drawn by four horses drove around its interior. Lackington claimed to have a stock of more than a million books. *The repository of arts, literature, commerce, manufactures, fashions and politics.* Vol. 1, April 1809, p. 251. B.L. C119.f.1.

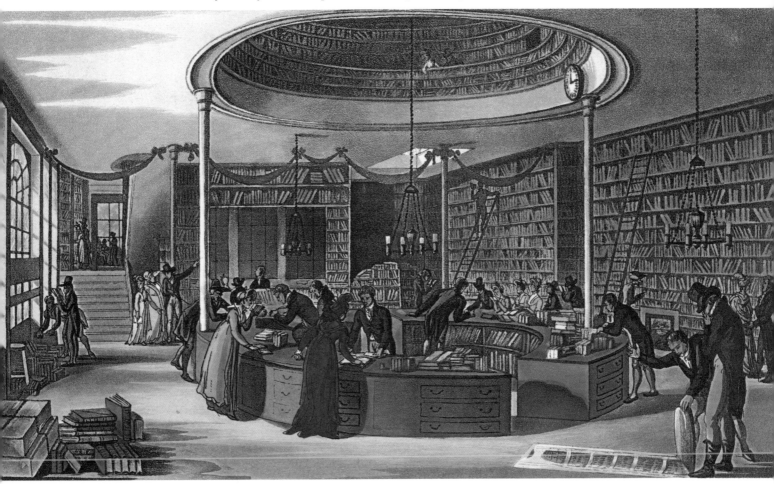

2 The discovery of the New World

Engraved title-page of Francisco Hernandez's *Rerum medicarum Novae Hispaniae*, Rome 1651.
The author had died before any of the records of his Mexican travels were printed.
An early example of posthumous publication. B.L.985.h.6.

The emergence of a scientific study of nature requiring a more systematic inventory of animals and plants coincided with the exploration of the Americas. Spanish and Dutch expeditions could often call upon the services of official or amateur artists to record new creatures and exotic vegetation. Colourful birds, especially parrots, attracted the eye, but plants possibly made a greater impact. They were shipped to Europe as seeds or living specimens in tubs of earth for botanists to study and for gardeners to grow for ornament or food. To Nicolas Monardes must go the honour of being the first to describe American plants in a book published in Seville in 1569, followed by a second part in 1571. An English translation, *Joyfull newes out of the new founde worlde, wherein is declared the rare and singular vertues of diverse … hearbes*, appeared in London in 1577.

In the year in which the second part of Monardes's book came out, a Spanish expedition sailed for New Spain (Mexico) under the leadership of Francisco Hernandez (*c.*1514–87), physician to Philip II of Spain. His mission was to enlighten the Council of the Indies in Seville on the population, antiquities and natural history of Spain's new colony. The detailed records and drawings accumulated by his team of observers and artists from 1571 to 1577, in the first official survey to be undertaken in the New World, were subsequently bound into sixteen folio volumes and stored in a wing of the Escorial in Madrid. Hernandez died before any of this material could be prepared for publication, but the King commanded Nardo Antonio de Recchi to edit it. He had extracted all the medical data before his death. Somehow a copy of de Recchi's manuscript reached America, where Father Francisco Ximenez, a lay brother at the Convent of Santo Domingo, translated it into Spanish and published it as the *Quatro libros* in Mexico City in 1615. It appears that de Recchi's notes formed the basis of an edition published in 1628 of which only a few copies were distributed. It has been conjectured that the sheets of this rarity with new preliminaries were used for another edition published in Rome in 1651: *Rerum medicarum Novae Hispaniae thesaurus, seu plantarum, animalium, mineralium, Mexicanorum historia ex Francisci Hernandez ....* Many of Hernandez's original records

Francisco Hernandez *Rerum medicarum Novae Hispaniae*, Rome 1651. Woodcuts of passion flower, sunflower, bison and manatee. The text provides Indian names with descriptions. B.L. 985.h.6.

were destroyed in a disastrous fire at the Escorial in 1671, and until recently it was believed that no other copies existed. Astonishingly, duplicates have been found in Spain, which deserve to be published as the pioneering investigation of America's fauna and flora by a scholar who has been called 'the third Pliny'.

The accession of Carlos III of Spain in 1759 brought to the throne an enlightened monarch, a patron of the arts and sciences who authorized the despatch of scientific missions to his colonies. In 1760 a young Spanish physician, José Celestino Mutis (1732–1808), accompanied the newly appointed Viceroy of New Granada (Colombia) to Bogotá, where he taught philosophy, mathematics and astronomy at the local college. Distracted by the wars in Europe, the King ignored repeated appeals from Mutis for money, equipment and books to carry out zoological and botanical research. Mutis left the botanic garden which he managed in Bogotá to superintend the royal mines in the Andes until 1770. At the same time, however, he personally financed and directed a team of botanists and artists. When another viceroy met him shortly after his arrival in 1782, Mutis had taken holy orders. Impressed by what Mutis had achieved through his own initiative, the viceroy appointed him director of a botanical survey, the King confirming him as 'First Botanist and Astronomer of the Botanical Expedition of Northern America'. Much more pertinent was the provision of adequate funds for staff and equipment. Similar expeditions were launched in the Spanish possessions in Peru and Mexico.

At his Bogotá headquarters Mutis created a large herbarium, a superb library, and a collection of nearly 7000 flower paintings executed by his artists. On his deathbed in

Flora de la Real Expedición Botánica del Nuevo Reino de Granada (1783–1816)
is now in course of publication in Spain. Vol. 10, plate 35. *Kefersteinia graminea*.
Copy of a watercolour of an orchid painted by José Manuel Martinez, one of the artists
on the expedition to Colombia. Biblioteca del Real Jardin Botánico, Madrid.

1808 he expressed a wish that his labours of 47 years should be published. In 1810 New Granada declared its independence, but after the Napoleonic wars Spain endeavoured to crush the rebellious colonists. Intellectuals were among the first victims. Members of the Expedición Botánica, naturalists and artists were shot and Mutis's herbarium and botanical drawings were crated and despatched to Madrid's botanical garden. There they still remain, but at long last in 1954 there appeared the first folio volume of a projected *Flora de la Real Expedición Botánica del Nuevo Reino de Granada* in 51 volumes with a selection of the drawings representing nearly 2800 plant species. At the time of writing more than 20 volumes have already been published.

The Spanish, Portuguese and French were already established in the New World when the British staked a claim to American territory. Led by Sir Walter Raleigh, a small band of settlers landed on Roanoke island in the Carolina Outer Banks in 1585. They were evacuated the following year, but in 1587 more colonists occupied the abandoned fort. The settlement included a scientific observer and also a talented artist, John White, who became governor of this short-lived colony. Sir Hans Sloane had copies made of White's competent drawings of Indians, vegetation, birds and fishes and a few of them were incorporated by Mark Catesby in his *Natural history of Carolina* (1729–47).

MARIA SIBYLLA MERIAN (1647–1717)

Portugal briefly lost Brazil to Holland. Its first Dutch governor, Count Johan Maurits of Nassau–Siegen, was a distinguished patron of the arts and sciences. During his term of office from 1637 to 1644 botanical and zoological gardens were laid out while a team of scientists and artists surveyed and recorded the country and its inhabitants. Count Maurits himself financed the publication in Amsterdam of the natural history investigations of the Dutch botanist Willem Pies (or Piso) and the German astronomer Georg Marcgrave, *De medicina Brasiliensi … et historiae rerum naturalium Brasiliae* (1648), the first full account of the natural history of any region in the New World.

When control of Brazil reverted to Portugal, Holland still retained a presence in South America in Surinam (Dutch Guiana). Under the treaty of Breda in 1667 Britain had ceded Surinam to Holland in exchange for other territory. From 1683 Dutch Surinam was governed by the Surinam Society, its shareholders being the United West India Company, the city of Amsterdam and the exceedingly wealthy Cornelius van Aersen van Sommelsdijck. The colony exported sugar, tobacco and cocoa and until Merian arrived there in 1699 no one showed any interest in its natural history.

Maria Sibylla Merian, born in Frankfurt on 4 April 1647, was brought up in a family of craftsmen. Her father, a topographical artist and a skilled engraver, died when she was only three years old. Her mother then married Jacob Marrel, a painter of traditional Dutch flower compositions who taught his stepdaughter to paint and engrave. In 1665, at the age of eighteen, she married Johann Andreas Graff, one of Marrel's apprentices, and three years later they moved to Nuremberg. Now a mother, she still pursued her childhood interest in insects, concentrating on the transformation of caterpillars into butterflies and

Maria Sibylla Merian sitting at her desk with a botanical book behind her.
Royal Botanic Gardens, Kew.

moths, drawing the plants they fed on and their successive stages from egg, caterpillar, and pupa to mature adult.

Her first book, *Neues Blumen Buch*, issued in parts between 1675 and 1680, recorded garden flowers in neat hand-coloured copper engravings. It was a convincing demonstration of her competence as a flower painter, although some of her flower studies and insects had been unashamedly copied from Nicolas Robert's *Diverses fleurs* (*c.*1660). Naturalist friends now urged her to illustrate the life cycle of Lepidoptera. She drew and engraved 100 plates for *Der Raupen wunderbare Verwandlung und sonderbare Blumennahrung* (1679–83). Every plate depicted in pleasing compositions caterpillars feeding on particular plants in flower or fruit. The Dutch artist and entomologist Johannes Goedaert, who had illustrated an account of the transformation of insects in *Metamorphosis naturalis* (1662–69), concentrated on the identification of insect species. Merian was the first author to associate insects with their host plants.

It is a reasonable assumption that it was the disintegration of her marriage in 1685 that led her to seek refuge with the austere Labadist community in Waltha Castle in West Friesland. Her half-brother, already living there, had persuaded Merian, together with her two daughters and her mother, to take this decisive step. The sect, formed by a French mystic, Jean de Labadie, disapproved of marriage between a Labadist, which she had become, and an outsider such as her husband. She reverted to her maiden name and refused to return to her husband, who eventually divorced her.

Waltha Castle housed cabinets of plants and insects collected by Labadists in Surinam. This first glimpse of tropical insects tempted Merian to new areas of research. After her mother's death in 1690 she left the Labadist community for Amsterdam, where the natural history collections of the President of the Dutch East India Company and other citizens reinforced her resolve to travel abroad at some future time to study insects in their natural habitats. In the meanwhile she maintained herself by teaching painting, by selling her pictures and by trading in insect specimens. Her elder daughter's marriage in 1692 to a former Labadist who had business connections with Surinam perhaps encouraged her to visit the colony. She organized and financed her own expedition by selling her paintings and collection of insects, and prudently made her will before sailing to Surinam with her younger daughter, Dorothea, now 21 years of age, in June 1699.

Surinam, which later divided into Dutch, British and French Guiana, was inhabited by indigenous Arawak Indians, African slaves, Portuguese Jews and Dutch Protestants. After two months' voyage Merian and her daughter reached the capital, Paramaribo, a town of about 500 wooden dwellings, where she rented a place with a garden for breeding insects. She acquired some slaves to hack paths in the rainforest. They and the Indians who collected specimens for her were more helpful than the Dutch plantation owners, who viewed her collecting and feeding caterpillars, all the time sketching them, as eccentric behaviour. One excursion of about 40 miles by canoe took Merian to a remote Labadist plantation, but normally she confined her investigations to cultivated ground. Her fact-finding forays ranged beyond entomology to Indian customs – the designs the Indians painted on their bodies and their dietary habits, for instance. She tentatively tasted their dishes and judged the flavours of wild fruits. As well as painting insects on parchment with her daughter's assistance, she pressed specimens or pickled them in brandy. Becoming very ill in Surinam's hostile climate, Merian wisely decided after 21 months there to return to Europe. On 18 June 1701 she and Dorothea boarded ship with their personal belongings, paintings, beetles and butterflies in spirits, bottles of snakes, a small alligator, lizards' eggs (which she hoped to hatch on the voyage), chrysalises poised to change, and many sheets of mounted insects.

Merian arrived back in Holland on 23 September 1701 at the end of a memorable adventure without the support of sponsorship or the protection afforded to the wives of officials and traders. Her priority was now to find employment, since she was unlikely to profit from this independent achievement. It has been claimed that she produced over 50 watercolours for Rumphius's *D'Amboinische Rariteitkamer* (1705). It is debatable whether they were ever used, but without doubt she subsequently hand-coloured at least one copy of the book, now in the Plantage Library of Amsterdam University. The fees she earned from this work went towards paying for the engraving of plates for her Surinam book, which friends had encouraged her to publish. Additional finances were raised by selling her Surinam specimens and, eventually, her original paintings.

She was now too old to undertake the formidable task of engraving the 60 projected plates herself. She managed three, but the remainder were done by a team of three

LEFT
Maria Sibylla Merian *Metamorphosis insectorum Surinamensium*,
Amsterdam 1719.
Engraved frontispiece showing the author studying specimens
brought to her. Behind her an arch leads to a tropical vista.
B.L.649.c.22.

BELOW
Maria Sibylla Merian *Metamorphosis insectorum
Surinamensium*, Amsterdam 1705. B.L.649.c.26.
LEFT Plate 17. Spiders hunting humming birds on a guava tree
(*Psidium guineense*). She found many large black spiders on these
particular trees, where they fed on the ants that climbed them.
Engraved by J.P. Sluyter?

RIGHT Plate 5. This large black caterpillar fed on cassava
(*Manihot esculenta*). Merian reported entire fields of the crop
destroyed by this pest. She added the snake purely for decoration.
Engraved by J. Mulder.

OPPOSITE
Unpublished watercolour by Maria Sibylla Merian,
presumably when she was in Surinam. Besides insects she also
studied lizards, amphibians and snakes. At one time she
contemplated using her drawings of them in a companion volume.
Her son-in-law in Surinam was willing to send her specimens.
Prints and Drawings Department, British Museum, London.

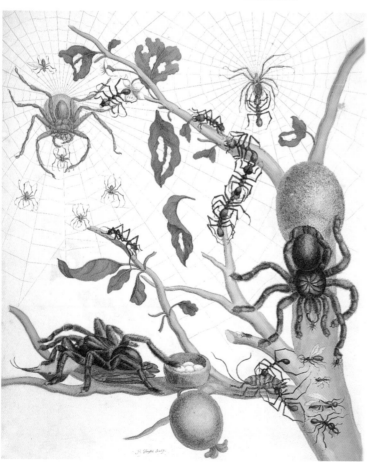

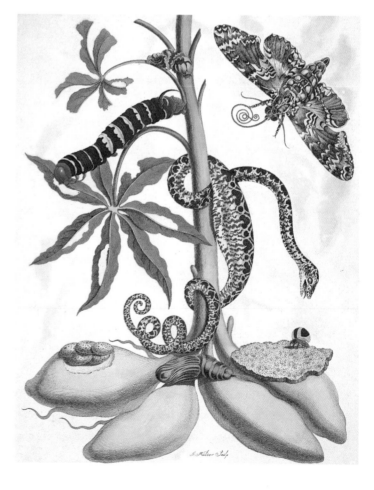

engravers. She had planned a German edition but abandoned it when friends could find her only twelve subscribers. She turned to James Petiver, an English botanist, to recruit English subscribers. Since several hundred copies were rather optimistically contemplated for the British market, an English translation was mooted. She sounded out Petiver on the feasibility of presenting a finely coloured copy to Queen Anne. Was there, she enquired, an English bookseller prepared to print the English text if she supplied the engravings? Petiver had no more luck than her German friends in getting enough subscribers. He himself was enthusiastic and had even planned to translate the work and to arrange the plates in a systematic order. His preliminary thoughts were included in his *Opera*, but no English version ever appeared.

Metamorphosis insectorum Surinamensium (Transformation of Surinam insects) was published in Amsterdam in 1705, in one version with a Latin text and another in Dutch. The prepublication price of fifteen guilders was increased to eighteen, and a hand-coloured copy by the author or her daughter cost 45 guilders. It is not known how many copies were printed, but six years later some were still available for sale. Almost certainly Merian lost money on this venture.

The book's 60 engraved plates portrayed about 90 studies of caterpillars through their evolution into insects. Most were drawn life-size. Merian added to her plants the local names given to them by Indians and European settlers, and Caspar Commelin from Amsterdam's Botanic Garden contributed Latin designations. Frequent observations on local uses of the plants enlivened her text.

Merian died before the second edition was brought out by the Amsterdam publisher J. Oosterwyk in 1719. He added another twelve plates, of which the first ten have been identified as Merian's own work. The remaining two may be by her daughter Johanna, known to be a competent artist. The engravings were used again in a French and Latin edition in 1726, a Dutch version in 1730 and finally in 1771 to accompany another French and Latin text. Nothing is known of the subsequent fate of the copper plates.

Merian produced several copies of her Surinam drawings, possibly as master copies for colourists or simply for sale. The English collector Richard Mead acquired 95 of her watercolours, which on his death in 1754 were purchased for George III (now part of the Royal Library collections at Windsor). Mead's contemporary, Sir Hans Sloane, also had a set of Merian drawings (now housed in the Prints and Drawings section of the British Museum). When Czar Peter the Great searched Holland in 1716 and 1717 for works of art he was advised by Georg Gsell, Dorothea Merian's husband. The sale to the Czar's representative of two large volumes of her paintings for 3000 guilders was agreed on the very day Maria Merian died. The Academy of Sciences in St Petersburg now boasts the best collection of Merian's drawings anywhere in the world.

As an artist Maria Merian was influenced by the technical expertise, the meticulous observation and the unerring composition of the Dutch school of flower painters. Her insects and flowers are integrated in rhythmic patterns, and vitality infuses her ecological vision. Her floral accessories are accurate portraits of tropical vegetation. She was a pioneer in a long line of women botanical artists and the first to depict insects in relation to their host plants. Perhaps her expedition to Surinam inspired later artists like Marianne North and Margaret Mee to travel abroad in order to paint plants in their natural surroundings.

MARK CATESBY (1682–1749)

Mark Catesby, who was privileged to see Merian's Surinam paintings, probably the set owned by Sir Hans Sloane, perhaps recalled her balanced compositions of creatures and vegetation when he illustrated his book on the natural history of part of eastern North America and the Bahamas.

Catesby, probably born in Castle Hedingham in Essex on 24 March 1682, was raised in Sudbury in neighbouring Suffolk. It is likely that his youthful enthusiasm for natural history was encouraged by the celebrated naturalist John Ray, who lived at Black Notley less than ten miles away. His uncle, Nicholas Jekyll, who knew Ray, may have introduced his nephew to the great man. One of Ray's friends, the Braintree apothecary Samuel Dale, befriended Catesby.

In 1712 Catesby went to Williamsburg in Virginia to stay with his sister Elizabeth, wife of William Cocke, Secretary to the British colony, having a 'passionate desire of viewing as well the Animal as Vegetable Productions in their Native Countries; which [were] Strangers to *England*'. He sent seeds to Henry Compton, Bishop of London, for his garden of American plants and to the London nurseryman Thomas Fairchild, and dried specimens of Virginian and Jamaican plants to Samuel Dale. When he returned in the autumn of 1719 he had little to show for his seven years abroad, but it was a seminal apprenticeship which enlarged his knowledge of exotic plants and animals, extended his collecting experience, tested his skill in drawing, and marked him out as a promising naturalist.

Through Samuel Dale, grateful for his protégé's presents, he was introduced to William Sherard, John Ray's successor as England's most eminent botanist. Their meeting was most opportune for Sherard, who saw the young man's drawings and learned that he was anxious to return to Virginia. John Lawson, Surveyor General of North Carolina, about to write an account of the natural history of the region, had been killed by Indians

Mark Catesby *A natural history of Carolina, Florida and Bahama Islands*, London 1729–47. Vol. 2, p.61.
Here Catesby used an engraving of *Magnolia grandiflora* by G.D. Ehret, whose signature
can be seen faintly on a leaf in the bottom left-hand corner of the print. B.L.687.l.3.

in 1711. In 1720 the artist and naturalist Eleazar Albin had declined an invitation to spend the summer painting in Carolina and the Caribbean. Catesby, available and eager to go, was offered an annual allowance of £20 by the Governor of South Carolina provided the Royal Society recommended him. As the Royal Society would not be financing his tour, Sherard had to find backers. Sir Hans Sloane, Charles Dubois, Richard Mead, the Duke of Chandos and Sherard himself formed a syndicate and Catesby left for Carolina in February 1722. Carolina was pioneering territory, lacking Virginia's stability, alert to hostile Indians and with suspicious Spanish colonists on its borders. In mid-June Catesby informed Sherard that he had purchased a 'negro boy' as a servant. On subsequent trips he had an Indian to carry his box of painting equipment. In difficult terrain he did his best to meet his patrons' requirements: despatching seeds in wooden boxes or gourds, snakes, small creatures preserved in bottles of alcohol, and boxes of birds, oven dried and sprinkled with tobacco dust. For some of his clients he sought plants likely to be hardy in Britain and with horticultural potential. He always avoided collecting more than once in the same place and at the same season of the year. But his compliance with their wishes stopped short of dispersing his drawings among them.

> *My designe was* [he wrote to Sloane] … *to keep my Drawings intire that I may get them graved, in order to give a general History of the Birds and other Animals, which to distribute Separately would wholly Frustrate that designe, and be of little value to those who would have so small fragments of the whole. Besides, as I must be obbligged to draw duplicates of whatever I send, that time will be lost which otherwise I might proceed in the designe and consequently be so much short in proportion to what is sent.*[5]

Catesby, who had collected extensively around Charleston and along the coastal plains by mid-1724, supported a proposal of Thomas Cooper, a Charleston physician, to mount an expedition to Mexico with him as collector. When Cooper's overture produced no positive response from London, Catesby informed Sir Hans Sloane in January 1725 that he was about to visit the Bahamas to paint its marine life, an excursion which added fishes, crabs, turtles and coral to his pile of drawings.

When he returned to London in 1726 friends encouraged him to publish his drawings. Unable to afford to employ engravers, he decided to do the work himself. Joseph Goupy, a French painter and engraver long resident in London, taught him how to etch copper plates. Catesby chose not to adopt the traditional method of cross-hatching to convey tonal values, preferring 'to follow the humour of the Feathers which is more laborious'.[6] Although Catesby supported himself by working in the London nurseries of Thomas Fairchild and Christopher Gray, he still needed financial assistance with the book. He found a benefactor in Peter Collinson, a Quaker woollen-draper with a fine garden at Peckham, who generously lent him large sums of money free of interest 'to

5. Charleston, Carolina, 15 August 1724. Sloane MS 4047, f.213. British Library.
6. *Natural history of Carolina*, vol. 1, preface, p.xii.

enable him to publish it, for the benefit of himself and family: else, through necessity, it must have fallen a prey to the booksellers'.[7] He was often Collinson's guest at meetings of the Royal Society, a source of potential purchasers of his forthcoming book.

In order to publicize the book, in 1729 Catesby issued a leaflet, his *Proposals* on its content and manner of sale. A part of his *Natural history of Carolina, Florida and Bahama Islands* would appear every four months, each with 20 coloured plates, priced at two guineas. Plain copies at one guinea were also promised, but since none is known to exist it is assumed that none was issued. Subscribers would not be expected to pay until 'each Sett is deliver'd; that so there may be no Ground to suspect any Fraud, as happens too often in the common way of Subscription'. He chose the part-issue method hoping, as he explained to the American botanist John Bartram, by this means to sell many more copies. Fellows of the Royal Society had an opportunity to inspect a copy of the first part at their meeting on 22 May 1729. Through publicity and recommendations, Catesby secured 154 subscribers of whom 29 were Fellows of the Royal Society.

His timetable for producing the book turned out to be an optimistic prediction. The tenth part came out in 1743, followed by an appendix in 1747. Not only did he engrave all the plates but he hand-coloured them as well, until he could afford to employ colourists working under his supervision. Modestly he hoped that his deficiencies as a painter were excusable and that his 'flat tho' exact manner' was acceptable. He painted flowers freshly picked, and living specimens of birds, fishes and reptiles.

The folio format of Catesby's book allowed him to draw many of his bird studies life-size. His first eight plates presented them in a conventional posture, perched on a tree stump or on the ground without any decorative background. Thereafter he broke with tradition by introducing fragments of the vegetation on which they fed. Only when he drew fish did he revert to solitary portraits. He usually created a compatible landscape but sometimes showed a flippant disregard for ecological authenticity by rather peculiar associations of flowers and creatures. A flamingo, for instance, is posed next to a grotesquely large piece of coral which looks like a stylized tree (vol. 1, plate 73). Frick and Stearns conjectured that Catesby disregarded harmony if it meant he could reduce the number of plates by crowding more subjects on them. David Wilson argues that he deliberately introduced disparate elements for aesthetic reasons. Catesby himself asserted that he observed environmental associations 'where it could be admitted of'.

Like his contemporaries he had no scruples about copying the work of other artists. From Sir Hans Sloane's collections he adapted seven drawings by John White and a sketch by Everhard Kick. Four drawings by Georg Dionysius Ehret, a young German botanical artist, feature in his second volume: *Magnolia grandiflora* (plate 61), *Magnolia tripetala* (plate 80), *Coccoloba uvifera* (plate 96) and *Magnolia virginiana* (plate 15 in the Appendix). He never acknowledged such indebtedness, except the use of White's swallow-tail butterfly. His Appendix included plants and animals he had not personally seen in

---

7. D. Turner *Extracts from … correspondence of R. Richardson*, 1835, p.401.

America. Bartram and other American naturalists sent him specimens; London nurseries provided flowers grown from American seed.

After 20 years' labour Catesby's great work was finished. He dedicated the first volume to Queen Caroline and the second to her daughter-in-law, Augusta Princess of Wales. He drew 171 plants, 113 birds and an assembly of fishes, insects and animals on 220 plates. Parallel columns of English and French text describe them.

The *Natural history of Carolina* was well received but there were inevitable critics. Linnaeus had reservations about its accuracy, although he based a number of new species on Catesby's plates. Thomas Jefferson considered them extravagantly coloured, and his fellow countryman Alexander Garden deplored their 'many blunders and gross misrepresentations'.

When Catesby died in 1749, aged 70, he left his widow all unsold copies. She reduced the price of each set by half a guinea to encourage sales; two years later she sold whatever remained, including the copper plates, for £400. It is not known whether this

Mark Catesby *A natural history of Carolina, Florida and Bahama Islands*, London 1729–47.
LEFT Vol. 1, p.17. Large red-crested woodpecker. Catesby depicted eight species of woodpeckers. Studies of birds dominated volume 1. B.L.687.l.3.
RIGHT A page of insects by Mark Catesby. He made no separate study of them in his book but decorated some of his plants with them. B.L. Add MS 5271, f.174.

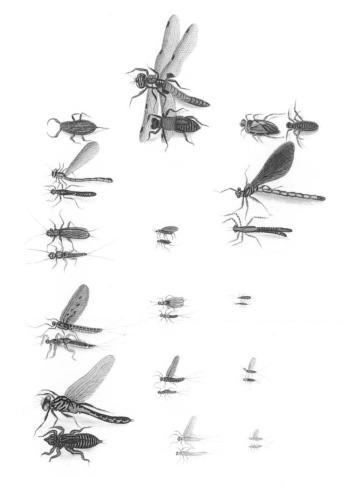

transaction included Catesby's original drawings, but in 1768 George III paid the London bookseller Thomas Cadell £120 for them. They are now in the Royal Library at Windsor.

Public interest in the *Natural history of Carolina* persisted. Smaller versions of eleven of its plates appeared in the *Gentleman's Magazine* between 1751 and 1753, seven of them in colour. Some were reproduced in Catesby's posthumously published *Hortus Britanno–Americanus* (1763). His friend the ornithologist George Edwards brought out new editions in 1748–56 and 1771. The book was popular in Germany, where one pirated version incorporated plates from two of Edwards's book as well. This Nuremberg firm also marketed a French edition between 1768 and 1776. A German edition of Buffon's *Histoire naturelle* copied many of Catesby's plates.

The *Natural history of Carolina* remained in demand for as long as it was the only book on North American natural history. Its author was neither a great naturalist nor a distinguished artist, but by placing his birds in an ecological setting he established a new trend in bird illustration. Frick and Stearns rightly dubbed him 'a colonial Audubon'.

Mark Catesby *A natural history of Carolina, Florida and the Bahama Islands*, London 1729–47.
LEFT Appendix, p.20. *Bison americanus* and *Robinia pseudoacacia*. Not satisfied with his own field sketch of a bison, Catesby adapted the drawing by Everhard Kick in Sir Hans Sloane's collection.
Royal Botanic Gardens, Kew.
RIGHT Vol. 2, p.70. Water frog and species of Sarracenia. B.L.687.l.3.

3 Investigating the flora of Asia

HENDRIK ADRIAAN VAN REEDE TOT DRAKENSTEIN
(1636–91)

Scientific and commercial information on the natural resources of the Indian sub-
continent, Ceylon and South-East Asia was gathered by personnel, often physicians,
of the two most powerful trading bodies in the seventeenth and eighteenth centuries: the
British East India Company and the Dutch Verenig de Oost-Indische Compagnie.

The latter – the United East India Company, formed in 1602 – wielded wide-ranging
powers granted by the Dutch Government. It could recruit personnel for its own army,
fortify its trading posts, conduct wars and, east of the Cape of Good Hope, negotiate
treaties. Rival settlements belonging to the Portuguese were seized, and before the close
of the seventeenth century the Company had a chain of strategically placed trading posts
extending from the Persian Gulf to India and Ceylon, Japan and the Moluccas in South-
East Asia. Spices, for which there was an insatiable demand, were especially prized, but
other crops were also sought. Dutch officials, encouraged to investigate the vegetation
of the Company's overseas possessions, compiled floras such as Van Reede's *Hortus Indicus
Malabaricus* (1678–93) and Rumphius's *Herbarium Amboinense* (1741–50).

The most ambitious and influential was undoubtedly the twelve-volume account
of the plants of Malabar. Now part of the modern state of Kerala in South-West India,
Malabar is a long, narrow coastal region, flanked by the mountains of the Western Ghats,
supporting a tropical rainforest and a fertile plain. To the Portuguese and the Dutch it
was a land of abundance: groves of coconuts and palms, teak and sandalwood, and the
coveted spices. Van Reede's desire to examine this luxuriance was influenced as much by
his own curiosity as by his obligations to the Company.

Hendrik Adriaan van Reede tot Drakenstein[8] enlisted in the Company's army and
served with distinction in five Dutch campaigns in Portuguese Ceylon and Malabar.
Impressed by his performance, Rijklof van Goens, the Governor of Ceylon, entrusted
him with several diplomatic missions. Promotion followed. Appointed First Captain and
Sergeant-Major of Ceylon in 1667, Van Reede commanded all the troops, including the
garrisons in Malabar. During the years he was Commander of Malabar (1660–77) he
defeated the Zamorin of Calicut, established good relations with the Rajah of Cochin
and deterred a French naval invasion. Malabar became administratively separate from
the Government of Ceylon, a development opposed by Van Goens, who was determined
to retain its subordinate status. Van Reede's unwillingness to defer led to a deterioration
in the friendship between the two men.

---

8. There are several versions of his name: Rheede or Reede; Draakestein, Drakestein, Drakenstein or Draekestin. The
form adopted by J. Heniger in his definitive biography has been chosen here.

Van Reede and other senior officers found their military operations impeded by a lack of fresh supplies of medicines for their troops. As drugs imported from Holland invariably lost their potency during long sea voyages to the East, the Company's Chief Medical Officer in Batavia (Jakarta) was currently investigating local plants as an alternative source of medicine. In 1669 he asked the Government of Ceylon to search for suitable plants.

On his own initiative Van Reede installed a laboratory in his house in Cochin. A chemist he engaged to extract oils from native plants succeeded in distilling oil from the roots of wild cinnamon. It is not known when this laboratory was set up, but it is conceivable that Father Matthew of St Joseph had advised him, perhaps even suggesting a programme of research. The two men had made contact in 1663 and met again during the early 1670s when Father Matthew treated some of Van Reede's sick personnel.

Father Matthew, whose baptismal name was Pietro Foglia, was born in southern Italy about 1617. He had studied medicine at Naples University and entered the Carmelite Order about 1639. A peripatetic life as a missionary began in Syria, took him to Persia, then to Gujarat in northern India and to Malabar in the south. He devoted his limited leisure time to botany, sketching the plants of the Mediterranean region and the lesser-known flora of Mozambique and India. A fragment of his accumulated notes and

Watercolour drawing of the talipot palm (*Corypha umbraculifera*).
When it was engraved for the *Hortus Indicus Malabaricus*, 1682, vol. 3, fig. 2, the figure of the European artist was omitted. The artists employed by Van Reede sometimes embellished their drawings with people and animals, but most of them were excluded from the engravings.
The British Library has 653 original drawings done for this book, B.L.Add MS 5029, no. 2.

H.A. van Reede tot Drakenstein *Hortus Indicus Malabaricus*, Amsterdam 1678–93. Vol. 1.
Engraved title-page. A woman, presumably symbolizing Indian botany, seated in front of a conservatory,
rake in hand and a pruning fork at her feet, being presented with plants. B.L.453.f.7.

drawings was published in 1675. His rather rudimentary Indian sketches, intended only
as an *aide-mémoire*, convinced Van Reede that they could form the basis of a survey of
Malabar's vegetation. The monk readily concurred, but his output never attained an
acceptable standard: his brushwork was poor and his pen and ink drawings lacked detail.
When the erudite botanist Paul Hermann, at that time stationed in Ceylon, visited Van
Reede, he judged Father Matthew's work unsuitable for publication. Although he was
no longer a principal collaborator, Van Reede later acknowledged him as the founder of
the *Hortus Indicus Malabaricus*.

Van Reede now turned to Indian physicians for professional assistance. On his behalf
they purchased medicinal plants from a local herbalist and organized collecting forays in
the neighbourhood of Cochin. Van Reede took advantage of his official tours in Malabar
to harvest more specimens. On one expedition he deployed his entire party of 200 men
as collectors. Over a period of two years the same plants were gathered in different seasons
to obtain both flowers and fruit. Three or four artists were kept busy painting them. Two

of them, one an ensign in the Company's army, came from families of artists in Antwerp and Utrecht. An advisory committee of 15 or 16 Indian physicians and scholars and Dutch doctors considered the specimens and the paintings. The Indians identified the plants by their vernacular names and provided botanical data, geographical distribution and local uses, based on their own knowledge and on materia medica manuscripts and traditional verse, the oral repository of Ayurvedic medicine. An official translator recorded their deliberations in Malayalam (the local language) and in Portuguese, which a Dutch clergyman, Johannes Casearius, rendered into a Latin text. In due course the first instalment was despatched to Holland.

But all was not going well for Van Reede. He had never enjoyed good relations with Gelmer Vosburg, his second-in-command, who criticized his administration in a report submitted in June 1676. Vosburg censured Father Matthew as a 'crafty Italian' priest who had ingratiated himself with Van Reede. He condemned Casearius for neglecting to convert Catholics to Protestantism. His antipapist polemics pleased Van Goens, now Director-General of India, at his headquarters in Batavia. Following a reprimand, Casearius offered to resign. In October 1675 a frustrated Van Reede informed the government in Holland that he wished to return home. Father Matthew's position worsened with the arrival of a Carmelite commission to replace him with another priest more acceptable to the local bishop. Van Reede, vigorously defending Father Matthew as a good friend of the Dutch, expelled the commission. Two years later, with Van Reede back in Europe, the commission returned to Malabar but was foiled in its attempt to kidnap Father Matthew and take him to Portuguese Goa. Eventually the elderly monk did leave Malabar; he is believed to have died in northern India.

The team of advisers and experts so assiduously built up by Van Reede dispersed when he and Casearius embarked for Batavia in March 1677. His companion died a few months later. One might have expected Van Reede to abandon his flora, but such was the determination and the resilience of the man that he found a substitute for Casearius to write a Latin text and a replacement botanist, the Dutch physician Willem ten Rhijne. These two persevered with the *Hortus* during the six months Van Reede stayed in Batavia. Before he sailed for Holland in October 1677 he took the precaution of leaving a duplicate set of the text and drawings with ten Rhijne in case his copy got lost on the voyage.

Arriving back in Holland in 1678, Van Reede was presented with a copy of the first volume of the *Hortus Indicus Malabaricus*, published a few months earlier. The notes and drawings he had despatched from Malabar had, without his knowledge, been edited by Arnold Syen, Professor of Botany at Leiden. Van Goens predictably ridiculed the book. Ceylon, he asserted, had better plants than Malabar. The second volume was already in preparation. Syen, having assumed that he had received the entire manuscript, had arranged with the publishers, Johannes van Someren and Jan van Dyck of Amsterdam, to produce two volumes of similar size. He allocated most of the trees to the first volume, with the remainder in the next along with shrubs and herbs. Van Reede had not wanted anything published until all the plants had been collected.

The death of Syen and both publishers in 1678 delayed the appearance of volume 2 until 1679. Syen had been replaced by Jan Commelin, a capable Amsterdam botanist. No doubt disappointing sales had deterred the widows of the deceased publishers from continuing with the work. When Van Reede persuaded Johannes Munnicks, Professor of Anatomy and Botany at Utrecht, to compile concise botanical descriptions, he pledged himself, Munnicks and Commelin to present the publishers, now enlarged by the participation of Hendrik Boom, with the manuscripts of two volumes a year. Notwithstanding this agreement, progress remained erratic. Munnicks lasted until volume 5 (1685), his reluctant successor, Theodorus Janssonius van Almeloveen, coped with volume 6 (1686), and Abraham van Poot, an Amsterdam physician, dealt with the remaining volumes, struggling to interpret all the notes he had inherited since Van Reede had returned to the East. Commelin diligently supplied commentaries, consulting, whenever necessary, the excellent resources of the Amsterdam Botanic Garden.

Van Reede never saw the twelfth and final volume, published in 1693. In December 1684 he had left Holland as Commissioner-General of the Western Quarters (of Asia). After a long voyage inspecting Dutch settlements, he sailed from Malabar in November 1691 for Surat in northern India. He died at sea off Bombay on 15 December, and was

H. A. van Reede tot Drakenstein *Hortus Indicus Malabaricus*, Amsterdam 1678–93. Vol. 1, fig. 1. Tenga (*Cocus nucifera*), drawn by Antoni Jacobsz Goetkint, an ensign in the Dutch East India Company and employed by Van Reede to record the plants of Malabar. Engraved by Bastiaan Stoopendael. Only two engravings were signed in the entire work; the other is fig. 39 in vol. 6, by Gonsalez Appelman. B.L.453.f.7.

H.A. van Reede tot Drakenstein *Hortus Indicus Malabaricus*, Amsterdam 1678–93.
Vol. 1, fig. 16. Ily (*Bambusa bambos*). Indians cooked and ate its seeds. B.L.453.f.7.

buried at Surat in a magnificent mausoleum which dominated all other tombs in the European cemetery.

His conception of the *Hortus Indicus Malabaricus* had been ambitious, its publication beset by difficulties but its completion a triumph. The work describes 729 plants, mostly from Malabar, some from other parts of India and Ceylon, with some commercial crops introduced from Japan, China and Malacca. Abraham van Poot translated the first two volumes into Dutch (*Malabaarse kruidhof* (1689)) at the request of booksellers. Presumably it failed to attract the general public, since unsold stock was reissued, with a new title-page, in 1720. The copper plates and remaining books were all sold in 1721.

It succinctly gives the distinguishing features of every plant: foliage, flowers, fruit and seed, trunk or stem, and roots. Their usage in agriculture, commerce or medicine is mentioned. Syen's pragmatic arrangement of the first two volumes was succeeded by Commelin's adoption of the classification scheme used by the English botanist John Ray. The plants' names are engraved on each plate in Roman, Malayalam, Sanskrit and Arabic scripts.

Most of the 791 copper engravings are double-paged in order to show the plants life-size. Branches and stems thrust boldly across the spread of facing pages; leaves are rhythmically disposed, some curiously distorted; dissections are scattered along the bottom edge. In austere outline and with minimal cross-hatching, the Dutch engravers, of whom only two are known by name, have imbued the original drawings with clarity and elegance.

Before the publication of the *Hortus* very little was known about Indian vegetation. Garcia da Orta's *Coloquios dos drogas he coucas mediçinais da India* (1563) and Christobal Acosta's *Tractado de las drogas, y medicinas de las Indias Orièntalis* (1578) were slight in substance, limited largely to useful and medicinal plants sold in Europe. Van Reede's book was the first regional flora of the Indian sub-continent and for very many years indispensable for the study of tropical botany. John Ray cited more than 80 per cent of its plants in his *Historia plantarum* (1686–1704). It yielded over 200 new genera and species for Linnaeus, who praised it in the fifth edition of his *Genera plantarum* (1754): 'I have not put my whole trust in any author, excepting the work *Hortus Elthamensis* by the very celebrated Dillenius and the work *Hortus Malabaricus* by the illustrious Van Rheede, having very firm conviction in their accurate data.' Elizabeth Blackwell borrowed from some of its plates for her *A curious herbal* (1737–39). So did John Hill in *A decade of curious and elegant trees and plants* (1773). Hill brought out a London edition of the first volume of *Hortus Indicus Malabaricus* with reduced plates in 1774, before his death in 1775. M.S. Merian frequently refers to the work in her book on Surinam insects. In his youth the distinguished botanical artist G. D. Ehret coloured a set of its plates for a German banker. The *Transactions of the Linnean Society* ran a rather tedious series of taxonomic commentaries on it between 1822 and 1835. Sir Joseph Hooker attempted to identify all its plants in his *Flora of British India* (1872–97). Had it not been for Van Reede's interrogation of his Indian scholars, a great deal of ethnobotanical knowledge might have been lost.

Van Reede – soldier, diplomat and administrator – conceived the work and recruited collaborators: European and Indian scholars, physicians, botanists and artists. He gave the project momentum, and kept it going when at times it would have been easier to admit defeat.

GEORGIUS EVERHARDUS RUMPHIUS (1627–1702)

Van Reede reached Batavia as an ordinary soldier in 1657, some four years after Rumphius, the author of the only work to rival the scope of the *Hortus Indicus Malabaricus*. It is doubtful whether the two men ever met, since Rumphius was stationed in the distant Moluccas.

Georg Everhard Rumf (as with Van Reede there are variant spellings of his name, but in this account the Latinized form by which he is now remembered will be used) was born in Hesse in Germany, the son of an engineer and builder. An opportunity to travel arose when a German nobleman recruited mercenaries on behalf of the Republic of Venice. Rumphius and the other volunteers were taken to a Dutch port, ostensibly for embarkation to the Mediterranean. They discovered too late that they were joining an expedition, under the aegis of the United West India Company, to protect Dutch settlers in Brazil under attack by the Portuguese. Fortunately for Rumphius they never reached

Brazil, where the Dutch were suffering heavy casualties. Their ship may have been intercepted by the Portuguese, for in 1646 he found himself in Portugal, where he remained for several years. On his return to Hesse in 1649 he found employment as a building supervisor. With his mother's death in 1651 he soon afterwards enlisted in the Dutch East Indies Company and sailed on Boxing Day 1652 to Batavia.

Batavia, the administrative headquarters of the Dutch possessions in Asia, tried to replicate the layout of a European town, with stone houses intersected by straight, tree-lined avenues. His fluency in German, Dutch and Portuguese and his building experience made Rumphius a valuable asset to the island of Ambon in the Moluccas, where he was to spend the rest of his life.

This strategically important island, just under 400 square miles in area and centre of the clove industry, is mountainous, with a high rainfall but with a more equable climate than that of Batavia. The Dutch had ejected the Portuguese who discovered it and massacred British traders who dared to establish a post there.

G.E. Rumphius *D'Amboinische Rariteitkamer*, Amsterdam 1705. B.L.459.d.7.
LEFT Frontispiece of crustacea and shells being brought to specialists for examination. Arches are liberally garlanded with shells, crabs, etc. Drawn by Jan Goeva and engraved by Jacobus de Later.
RIGHT Portrait of Rumphius by his son, P.A. Rumphius, showing his blind father carefully feeling specimens on the table. Engraved by Jacobus de Later.

G.E. Rumphius *D'Amboinische Rariteitkamer*, Amsterdam 1705. B.L. 459.d.7.
LEFT Plate 6. Sea crabs (*Cancer* sp.). The author reported that not all the crabs appeared to be edible.
RIGHT Plate 23. Collection of shells. Much of the text and many of the illustrations are devoted to shells.

Rumphius spent three years at Larike, a small coastal settlement, as a junior merchant in the civilian arm of the Company. Eminently self-reliant, his isolation gave him the opportunity to explore the natural world around him, gradually, one supposes, considering the possibility of writing about it. His reputation as a scholarly official who could speak several native languages guaranteed a sympathetic response when he asked the directors of the Company in Amsterdam for books and scientific instruments to assist him in writing a book on the animals and plants he was observing and collecting in the East Indies. His official duties were reduced to give him more time for research, provided the Company's interests did not suffer.

He lived with Susanna, probably a native woman, by whom he had a son, Paulus Augustus, and at least two daughters. This very pleasant existence – a congenial companion who shared his interest in the island's flora, and his absorbing work – abruptly came to an end when, at the age of 42, Rumphius became totally blind through glaucoma. He blamed the strong sunlight, but long hours studying by candlelight may have contributed to this disaster. He was forced to relinquish his post and move to the capital city of Ambon, with, for the time being, no loss of salary.

Despite this appalling handicap he carried on as best he could with his account of Ambon's flora. He now dictated to an amanuensis, but being no longer able to continue his Latin text since there was no classicist on the island, he started afresh in Dutch. He consoled himself with the thought that Dutch might make the work more widely accessible. Others added to the coloured illustrations he had already accumulated. Catastrophe struck again in February 1674 when a violent earthquake killed Susanna and one of his daughters. In 1682, for reasons that are not clear, he sold his collection of natural curiosities, mainly shells, painstakingly collected over 28 years, to Cosimo III, the Grand Duke of Tuscany. He lost the rest of his specimens, his library, drawings and papers in a fire which destroyed the European quarter of the city of Ambon in 1687. Two years later his second wife died. Sixty-one of the new illustrations he had made were stolen in 1695.

Rumphius could never have survived these afflictions without the support of friends and sympathetic officials. Clerical assistance was provided by the local Governor. The Governor-General from 1679–81, Rijklof van Goens (Van Reede's superior in Ceylon), found him some suitable assistants. His son Paulus Augustus, on his return from Holland in 1685, served both as secretary and artist until it interfered with Company duties. Before he gave up helping his father he instructed a draughtsman, Philip van Eyck, who, in due course, trained Pieter de Ruyter. The last artist was a professional, Cornelius Abramsen. In 1690 six of the projected twelve parts of the *Amboinische Kruid-boek* were sent to Batavia for shipment to Holland. Providentially, Johannes Camphuys, the Governor-General and an enthusiastic naturalist, had them copied before despatch. In 1692 the Dutch ship carrying them was sunk by a French naval squadron and the original manuscripts were lost. Camphuys ordered another copy to be made, giving Rumphius the chance to make any amendments to the text and illustrations. The twelve parts of the *Kruid-boek* eventually reached Amsterdam in August 1697. Rumphius compiled a supplement, or *Auctuarium*, which joined them in 1704.

The Company's directors applauded Rumphius's industry, welcomed his *Kruid-boek* as 'a praiseworthy work', rewarded him by promoting his son to the rank of merchant, but refused to publish the book, fearing it might contain information of commercial benefit to their competitors. Two years later they relented, but only on condition that the work be resubmitted for vetting. With such a discouraging requirement, the manuscripts languished in one of the Company's warehouses for almost 40 years. This procrastination no longer affected Rumphius, who died on Ambon in June 1702, aged 74.

J. Burman, Professor of Botany at Amsterdam University, rescued Rumphius's manuscript from neglect when he volunteered to edit it for publication in 1739. His Latin translation together with Rumphius's Dutch text appeared in parallel columns in six folio volumes issued by a consortium of publishers between 1741 and 1750, followed by the *Auctuarium* in 1755. *Het Amboinische Kruid-boek* (The Ambonese herbal) – commonly referred to by its Latin title, the *Herbarium Amboinense* – occasionally reached beyond

G.E. Rumphius *Herbarium Amboinense*, Amsterdam 1741–50. Vol. 1, plate 8.
Corypha rotundifolia, now *Livistona rotundifolia*. B.L. 39.h.1.

Ambon to Banda, Java and other islands in the Moluccas in its phytogeographical coverage. About 1200 plant species are botanically described and their uses noted; their names
are given in Latin, Dutch, Ambonese and Malay, and occasionally in Portuguese, Hindi
and Chinese; 695 plain engravings accompany the text. There is no record of the number
of copies printed, but 500 has been suggested. A second edition was published in 1750.

Even today no serious research on the flora of South-East Asia can ignore the book.
Professor E.D. Merrill of the Manila Herbarium was the first botanist to identify most
of Rumphius's species.[9] A young American botanist on his staff, C.B. Robinson, spent six
months gathering specimens and data in Ambon. His fieldwork was cut short when he
was murdered in a remote part of the island in December 1913. His death was attributed
to a local superstition that every year during the months of November and December
strange people were abroad looking for victims to decapitate. Robinson, who habitually

9. E.D. Merrill *Interpretation of Rumphius's herbarium Amboinense*, 1917.

dressed in khaki, wore a felt hat and carried a hunting knife, looking unlike other white people the natives had encountered, was believed to be one of them.

Rumphius never saw the publication of his two major works on Ambon. He himself rated the *Kruid-boek* as the better of the two, but the general reader might prefer the more anecdotal *D'Amboinische Rariteitkamer* (The Ambonese curiosity cabinet), published in 1705, with its intimate knowledge of the habitats and behaviour of marine creatures and the author's musings on disparate topics – the origin of amber, for example, or the extra-ordinary stones and hard objects excreted by animals or found in their stomachs.

Rumphius embraced Nature with his senses as well as with his intellect. His aware-ness of ecological factors placed him well ahead of his time. H.L. Strack, who led the Rumphius Biohistorical Expedition to Ambon in 1990, was impressed by his accuracy and reliability. E.D. Merrill, who devoted years to studying him, recognized him as 'one of the outstanding naturalists of all time'.

Pen and ink sketch of William Roxburgh in old age by W.J. Hooker.
Royal Botanic Gardens, Kew.

WILLIAM ROXBURGH (1751–1815)

The newly-formed British East India Company's first mission to India in 1608 sought trading facilities from the Moghul Emperor Jahangir, who granted it four centres of operation in northern India. After its ignominious expulsion by the Dutch from South-East Asia in 1623 it intensified its activities in India. By the 1640s it had established more than 20 trading posts or factories and settlements. A first-rate harbour became available in 1665 when Bombay passed to Britain as part of the dowry of Charles II's Portuguese queen.

For years reports and rumours of the wealth of India's natural resources had aroused the curiosity of botanists and traders. Van Reede's *Hortus Indicus Malabaricus* conjured up a vision of a botanical paradise for John Ray: 'Who would believe that in the single province of Malabar, and that not so very large, there are found more than three hundred wild trees and shrubs, and probably many more?'[10] Yet such books, commanding respect with their splendid illustrations, were of limited value until Linnaeus made the identifying and naming of plants relatively simple. Paul Hermann's listing of Ceylon's plants in the *Flora Zeylanica* (1747) was the first tropical flora to adopt Linnaeus's scheme. Those of his students who went overseas, trained in the new methodology, became his 'apostles'.

John Gerard König (1728–85) was Linnaeus's advocate in India. An appointment as surgeon and naturalist in the Danish mission at Tranquebar had brought him to South India in 1768. An attractive salary tempted him to serve the Nawab of Arcot as his naturalist. Better facilities for research and opportunities for travel brought him to Madras as the East India Company's naturalist in 1778. There he met other Company officials with similar interests, among them William Roxburgh, who joined him on field trips. König also botanized with the physician Patrick Russell, the two of them spending evenings together annotating their copies of Van Reede and Rumphius. Russell advised him to submit a selection of his notes and drawings of useful plants to the Court of Directors of the East India Company as evidence of his profitable research. Nothing came of Russell's proposal in 1784 to Sir Joseph Banks, the influential President of the Royal Society, that the Directors might consider financing the publication of König's work.

When König died of dysentery in 1785 the Council of the Madras Presidency appointed Russell as his successor as naturalist. His friend's premature death reinforced Russell's resolve that his efforts should not go unrecognized. He reminded Banks of his earlier request that König's papers be published. At the same time he set about convincing the Governor of Madras and the local Medical Board of the desirability of 'a work limited to the useful plants of Coromandel which, though less generally interesting to the botanists in Europe … might prove of real service to India'.[11] (Coromandel was the southern part of India's eastern seaboard.) Russell proposed that König's papers be supplemented by contributions from himself and local naturalists. He envisaged the first number containing descriptions of 30 or 40 plants of 'established utility'; three more numbers would complete the work. He nominated Sir Joseph Banks as the most suitable person to oversee its publication, which the East India Company might be persuaded to subsidize.

The Court of Directors in London, while sympathetic to the project, naturally required a detailed statement of costs. Russell, ignorant of the economics of printing and publishing, referred them to Banks, who supplied calculations and his recommendations towards the end of 1788.[12] He believed that 20 plates should illustrate each part rather than Russell's preference for 50. William Curtis's elegant *Flora Londinensis*, then being published, could serve as a model. James Sowerby, a botanical artist and engraver who supplied costs for the hand-colouring of the plates, was keen to get the commission. Russell would be responsible for despatching dried specimens, coloured drawings and descriptive notes. Banks undertook to find an editor and to supervise publication. He

10. C. E. Raven *John Ray: naturalist*, 1950, p.231.
11. W. Roxburgh *Plants of the coast of Coromandel*, vol. 1, 1795, p.iv.
12. 25 November 1788, Dawson Turner transcripts of Banks's letters, vol. 6, ff.91–93. Natural History Museum, London.

felt confident that if the entire edition were sold it would yield a modest profit for the Company. With that assurance the Court of Directors informed the Madras Presidency of their 'readiness, at all times … to promote the improvement of Natural History' and of their 'approbation of Dr Russell's proposal for publishing a select collection of useful Indian plants'. Russell, however, had left India in 1789 just before this letter reached Madras, and it was left to his successor, William Roxburgh, to implement the Directors' wishes.

William Roxburgh had entered the services of the East India Company as a surgeon's mate on an East Indiaman and was selected for the post of assistant surgeon at the Madras General Hospital in 1776. Promotion to full surgeon in 1780 took him to the garrison station at Samalkot some 200 miles north of Madras. There he was allowed to grow spices, coffee, breadfruit and other crops in an experimental garden. He still found time for botanical studies, supervising two Indian artists who drew the plants he collected.

He was an obvious candidate for Patrick Russell's vacant post, to which he was appointed in May 1790. In September he despatched his first consignment of drawings and notes to London. In December he told Banks that he had nearly 700 drawings ready, a third of them grasses. Banks's reservations about their quality clearly annoyed him: 'It would have given me much satisfaction if you had mentioned what the defects were that my drawings and descriptions had. I would then, probably, have been able to rectify them in those that are still to finish, but you have left me in the dark.'[13] He begged Banks not to publish them until they had been corrected or replaced: 'I hoped I am possessed of foresight sufficient not to expose my ignorance to the world.' Roxburgh felt Banks's criticisms so keenly that he even wrote to James Edward Smith, President of the Linnean Society of London, for an honest opinion.

His flagging confidence in his own abilities must surely have been boosted by his installation as Superintendent of the Botanic Garden in Calcutta in November 1793. The garden, originally conceived as a place where sago and date palms could be propagated and distributed throughout British India as an alternative source of food in times of severe famine, now grew many commercially viable crops. Banks, who advised on its development, saw it existing for 'the promotion of public utility and science'.

Roxburgh's duties as Superintendent went in tandem with his commitment to Banks and the embryonic book. By 1794 Banks had examined about 500 drawings executed by a small team of Indian artists, each one producing about twelve plant portraits a month. Roxburgh had recruited them from artists who, now deprived of the patronage of the disintegrating Moghul empire, found alternative employment as draughtsmen to the Company engineers and surveyors or received private commissions to produce pictorial souvenirs. His men had to abandon the style of fastidious miniatures associated with traditional Indian art and learn how to paint flowers, usually life-size, in a European idiom acceptable to botanists. Foliage, flowers and fruit had to be morphologically

13. 17 August 1792, Add MS 33979, ff.1761–73. British Library.

correct; floral dissections, vital for identification purposes, were mandatory. Though these artists adapted reasonably well to this scientific discipline, their work often displayed a geometrical stiffness, an urge to create decorative patterns. Nevertheless, they satisfied that censorious critic Sir Joseph Banks, who praised 'the accuracy with which the parts illustrative of the sexual system [of Linnaeus] are delineated'.

That commendation appeared in a letter he wrote to the Court of Directors in July 1794 about progress. He even asserted that the work of Roxburgh's artists was superior to that in the *Hortus Indicus Malabaricus* and the *Herbarium Amboinense*. Almost as an afterthought, he added that Van Reede's book now fetched more than five times its original price, a casual snippet of information clearly intended to reassure any hesitant or doubtful Director. He submitted samples of both plain and coloured plates. The former, besides being cheaper, allowed purchasers to colour them according to their own tastes. Rather than wait until all the plates had been engraved, Banks recommended publication in parts, a prudent method that would test public response at a modest outlay. Two parts, each with 25 plates, would be published annually. With a print run of 500 copies, he calculated, the unit cost of each part would be fifteen shillings, for which booksellers would pay eighteen shillings, retailing it at one guinea. The public, he was confident,

Sepia wash drawing by W.H. Fitch, based on a sketch by J.D. Hooker,
of W. Roxburgh's official residence in the Botanic Garden in Calcutta. Royal Botanic Gardens, Kew.

would regard such a price as reasonable, and the Company would profit by £75 from the sale of all the copies of each part. Having enjoyed good relations with George Nicol, the King's bookseller, during the publication of James Cook's *Voyage to the Pacific Ocean*, Banks recommended him as publisher. He ended his letter with a rhetorical flourish, stressing the benefits of such a book to 'the extension of the commerce of the Company & the improvement of our materia medica.'[14]

The reply of the Court of Directors was prompt and positive. George Nicol received an advance of £300. The first part appeared in May 1795, entitled *Plants of the coast of Coromandel*, 'under the direction of Sir Joseph Banks, P.R.S.', printed by William Bulmer. Of the hundred copies at the Company's disposal, a coloured one was sent to Roxburgh as a token of the Directors' 'approbation of his services'. In the event, Roxburgh never got his coloured copy, which was lost at sea, and had to be content with one of the 40 plain copies despatched to India for sale at 20 rupees each.

As soon as Roxburgh learned that the drawings were about to be published, he set about organizing and describing an accumulation of another 1100. Banks complimented him on the improvement in his artists' work: 'Many are charming. The grasses in particular are valuable in the highest degree to botanists as the fructification is delineated with accuracy.'[15]

The first two parts were well received, but after the fourth Banks admitted that sales were disappointing:

> *no books can sell on account of the pressure of Taxes & voluntary contributions;*
> *I cannot hurry on the bookseller tho I have tried, all I can therefore say is that the*
> *work is not stopped & that it will go on again at its usual speed of 2 fasciculi a year*
> *as soon as ever times change for the better.*[16]

The potential sale of a number of coloured copies had been lost when the Court of Directors insisted on presenting many of them to friends. Despite the recession, Banks assured Roxburgh that there would be no delay in printing the rest of the book.

Having paid George Nicol over £2000 up to 1799, the Court of Directors demanded a statement of sales. Nicol, unable to give any exact figures, insisted that 'he had sent the numbers to the most extensive channels of sale'. Not one of the 40 copies shipped to India had been purchased, despite advertisements in local newspapers, and only five copies of a companion work on snakes of the coast of Coromandel by Patrick Russell, also subsidized by the East India Company, had been sold. When six parts had been published with two more in the course of printing, and with receipts from sales amounting to only £400, the Company debated in 1801 whether to continue. The Napoleonic wars had disrupted sales on the Continent, where it was 'held in the greatest estimation', a temporary setback which would be resolved with the restoration of peace. Regardless of

14. 4 July 1794, Dawson Turner transcripts of Banks's letters, vol. 9, ff.52–56. Natural History Museum, London.
15. 29 May 1796, Add MS 33980, ff.65–66. British Library.
16. 9 August 1798, Add MS 33980, ff.59–60. British Library.

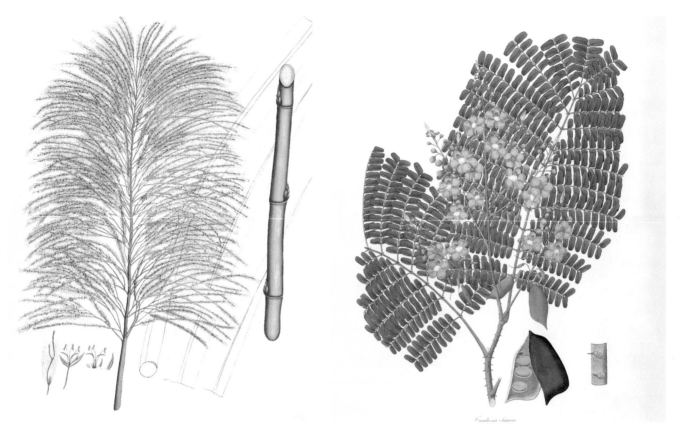

William Roxburgh *Plants of the coast of Coromandel*, London 1795–1820.
LEFT Vol. 3, plate 232. *Saccharum sinense*. According to Roxburgh this plant was introduced from China
to the Botanic Garden at Calcutta in 1796. It promised to be superior to Indian sugar cane. B.L.458.h.22.
RIGHT Vol. 1, plate 16. *Caesalpinia sappan*. Rated as a 'very valuable tree' by Roxburgh, it was
used as a source of red dye for cloth and wool, and also for sturdy fencing. B.L.458.h.2c.

OPPOSITE
Vol. 3, plate 234. *Trapa bispinosa*, now *Trapa natans* var. *bispinosa*.
A common aquatic plant with edible seeds. B.L.458.h.22.

whether the book made a profit or a loss, the Committee of Warehouses that was responsible for the Company's publications recommended its continuance, since it was 'an inducement to the Company's servants to devote a portion of their time and attention to scientific researches'.[17]

In 1810 its somewhat tardy progress was interrupted by the death of its editor, Jonas Dryander, and the loss at sea of a large consignment of drawings and associated notes. The ninth part, the first instalment of volume 3, appeared in July 1811. The appearance of the next part four years later coincided with the death of Roxburgh, and the Court of Directors consulted Banks about possibly ceasing publication altogether. The absence of constant and loyal subscribers is confirmed by the sale of individual parts, progressively declining from 145 for part 1 (1795) to 34 for part 9 (1811). 'It always depended on chance sale,' Nicol explained to the Directors. Surprised that no copies had been sold in Germany, 'that great mart for books of Natural History', he nevertheless urged the

17. 10 June 1801, MS Eur. F.25, ff.75–97. India Office Records, British Library.

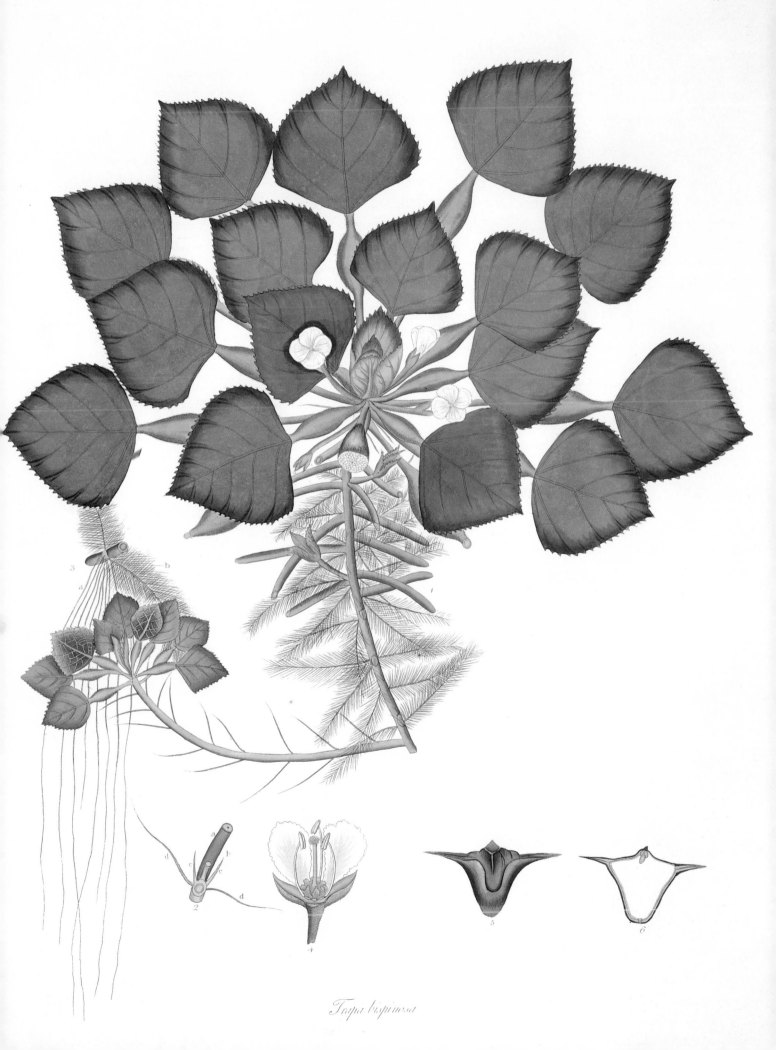

Trapa bispinosa

Directors to authorize the printing of two more parts, making twelve in all, to complete three magnificent volumes 'which will eventually do much more than repay the Directors their original advances, for this splendid work which will for ever form a monument of their liberality in allowing the work to be sold at prime cost'.[18]

The Court of Directors agreed to its completion, and in August 1816 Nicol announced that the last part should be ready by the end of the year. He exonerated himself from any responsibility for its protracted publication: 'But the fact is that some gentleman of the Committee [i.e. T. T. Metcalfe, a Director who had died in 1812] who managed that business, frequently threw cold water upon it, which was the true cause of the delay.'[19] The last part, which had been optimistically forecast for 1816, eventually emerged in March 1820. Nicol held the letterpress of about 150 to 160 copies of each part, but no plates, which were only printed whenever needed. He blamed the long period of gestation, nearly a quarter of a century, for poor sales. Nevertheless he was confident that sales

> *will be very considerable, both at home and abroad. ... It is far superior to the Hortus Indicus Malabaricus which has so long done honour to the Dutch East India Company ... but this I am certain of, that both in beauty & science the English work is greatly superior to the Dutch.*[20]

Sir Joseph Banks lived just long enough to see the last number. Without his recommendation it can be assumed that *Plants of the coast of Coromandel* would never have been approved, and without his participation it might have foundered after a few parts. Under his direction 300 drawings from more than 2500 submitted by Roxburgh were selected for engraving. He probably nominated Daniel Mackenzie, who had worked for him on other books, as principal engraver. He produced 250 plates, the rest being executed by Sansom, Girton, Peake and Weddell. All resisted any temptation to adapt or modify the drawings of the anonymous Indian artists. The stiffness and formality of the drawings have a certain ornamental charm. The hand-colouring of the engravings in transparent washes is of variable quality in the first two volumes but improves in the third. On one occasion Nicol apologized to the Company for the delay in producing the required number of hand-coloured copies because 'the woman who has coloured them so long and so well, has lately, by an accident' lost two of her best colourists.

The chief attraction of this folio work is undoubtedly the plates, but Roxburgh's contemporaries were also interested in his ethnobotanical notes. In the very first part he expended several thousand words on the cultivation and methods of dyeing cotton cloth with the roots of *Oldenlandia umbellata* (now *Hedyotis umbellata*), or chay root, then extensively grown in Coromandel. He discussed at length the sequence of dipping cloth in a powdered chay root solution, then into a mixture of ground *Terminalia* nuts and buffalo milk, followed by a solution of alum and turmeric, each stage concluded by a routine washing in cold water. The characteristic bright red colour emerged after a vigorous washing with soap. No other plant in the book received such a comprehensive account of its processing. Roxburgh usually confined himself to succinct statements:

woods suitable for shipbuilding, furniture, decorative carvings (a particularly hard one was recommended for scientific instruments); edible flowers, leaves and seeds; medicinal plants; flowers for tropical gardens. Some plants had no known commercial application but were admitted as being new to botany.

Plants of the coast of Coromandel is a worthy memorial to the patronage of the East India Company which played a seminal role in promoting the serious study of India's fauna and flora. Patrick Russell, who conceived the work in the first place, praised it as 'an example for the encouragement of the Company's servants abroad, to dedicate their leisure hours to useful research'.

The set of Roxburgh's Indian flower drawings which had been retained by the East India Company was transferred to the Royal Botanic Gardens at Kew in 1858. However, when the India Office, the successor to the Company, offered Kew 194 of the original copper plates in 1881, they were firmly declined:

> *Looking at the extreme imperfection of the series and the antiquated and cumbrous character of the work which has been practically superseded by the more recent publications of Wight and Beddome, Sir Joseph Hooker* [Director at Kew Gardens] *while thanking you for bearing this establishment in mind, is of the opinion that the only value of the plates at the present day is for disposal as old copper.*[21]

18. 31 May 1815, MS Eur. F.25, ff.107–08. India Office Records, British Library.
19. 28 August 1816, MS Eur. F.25, ff.169–71. India Office Records, British Library.
20. 17 March 1820, MS Eur. F.25, ff.119–21. India Office Records, British Library.
21. 29 August 1881, Herbarium donations to 1900, vol. 2, f.654. Royal Botanic Gardens, Kew.

4 A garden at Eichstätt

The voyages of Portuguese navigators during the fifteenth century took them to Madeira, the Azores, the Cape Verde Islands, down the west coast of Africa and across the Indian Ocean to the Indian sub-continent. Before the end of the century the Spanish had reached North America. Occasionally these adventurers brought back a random collection of plants and seeds, the first glimpse Europeans had of a new flora eagerly coveted by individuals and botanical gardens.

In the intellectual climate of the Italian Renaissance a new sort of garden emerged when universities set aside small plots of land for medicinal herbs or simples. Pisa led the way in 1543; Padua and Florence followed two years later; this fashionable garden reached Pavia in 1558 and Bologna in 1568. These botanical gardens provided specimens for demonstration and study in schools of medicine, displayed in related groups – poisonous, scented, bulbous and so on. They were described, named and their known or attributed properties investigated. Until about the middle of the sixteenth century most plants came from Europe and the Near East (Asia Minor, Egypt, Crete and Cyprus). A revival in botanical studies created a demand for plants in general, not confined to those of use to doctors. Botanical gardens now competed for new species brought back by traders and travellers. Through the agency of the Dutch East India Company the vegetation of the East Indies found its way to these university gardens. The New World was more accessible and soon Padua could boast possession of the potato, agave and sunflower. During the seventeenth century American species dominated the importation of plants.

The botanical garden attached to Leiden University, founded in 1587 for the use of the medical faculty, became an important centre in this international transit of plants. It achieved this eminence under the directorship of Charles de l'Ecluse (better known as Carolus Clusius), who had studied medicine in France and had translated into Latin Garcia da Orta's book on Indian plants and that of Nicolas Monardes on American vegetation. He had created a Hortus Medicus in Vienna before moving to Leiden. The foreign bulbs and seeds he received through a network of correspondents were generously distributed to friends, amongst them Joachim Camerarius the Younger. Both men advised or influenced the Bishop of Eichstätt in the creation of a garden soon to become one of the most celebrated in Germany. It has long since disappeared, but it is still remembered through a magnificent book, the *Hortus Eystettensis*, which described its choicest flowers.

The town of Eichstätt on the river Altmühl lies between Munich and Nuremberg. A castle which crowned a nearby hill, the residence of the bishops of the diocese of Eichstätt, had been built in the fourteenth century and subsequently altered and enlarged. Prince-Bishop Martin von Schaumberg rebuilt its eastern façade and began forming a

Basil Besler *Hortus Eystettensis*, Altdorf 1613.
Title-page. God introducing Adam to the Garden of Eden, flanked by King Solomon, who was reputedly interested in plants, and Cyrus, who created a famous garden in Persia. Surmounting the arch are the coat of arms of Prince-Bishop Johann Conrad von Gemmingen and the reclining figures of Ceres and Flora. The potted opuntia and agave herald the growing impact of American plants. Engraved by Wolfgang Kilian and hand-coloured by Georg Mack, whose name appears on the plinths supporting the opuntia and agave. The British Library's copy is one of the finest in existence. B.L.10.Tab.29.

garden. Walls were raised, paths and steps laid, summerhouses built, fountains overhauled and flower borders planted. Its development continued under Johann Conrad von Gemmingen, elected Prince-Bishop in 1595. A cultured cleric and a discriminating collector, he spent prodigiously on improvements to his episcopal palace. The building itself was reorientated to the west, facing the town below. It is, however, the transformation of its surroundings that concern us. Reports on improvements to the garden come from Phillip Hainhofer, an Augsburg aristocrat and art dealer, who sought *objets d'art* for German collectors including Wilhelm V, Duke of Bavaria. It was on the Duke's behalf that he presented himself to the Prince-Bishop of Eichstätt in May 1611. As the Duke's envoy his mission was to obtain copies of drawings in the Prince-Bishop's collection.

He was conducted around the palace's eight gardens, each filled with flowers 'from a different country', and admired displays of roses, lilies and tulips. A flower-painted panelled staircase led to the Prince-Bishop's quarters, where balconies were filled with pots and tubs of flowers. The rocky hill was in the process of being contoured, fertile soil brought up from the valley below and a system of irrigation pipes laid. The grounds took advantage of views of the surrounding countryside. Many of the plants were acquired from Brussels, Amsterdam, Antwerp and other places in the Low Countries. Possibly Clusius and Camerarius the Younger were sources of supply.

The bibliographical complexities of the book about this garden have been painstakingly investigated by Nicolas Barker. During his researches in German archives he discovered nothing to confirm absolutely Camerarius's involvement in the planning of the Eichstätt garden, but it can be argued that he influenced the presentation of the book describing its flowers. Joachim Camerarius the Younger (1533–98) tended a well-established garden in Nuremberg and wrote a *Hortus medicus et philosophicus* (1588). If he had collaborated with the Prince-Bishop it could only have been for a short while. What we do know is that after his death in 1598 the Nuremberg apothecary and collector of natural curiosities Basilius Besler (1561–1629) managed the garden and planned the book.

When Hainhofer visited Eichstätt in 1611 the Prince-Bishop regretted not being able to show his guest some of his best flower paintings, which were in Nuremberg. Besler had persuaded him to have them engraved there for a book which would be dedicated to his employer. Besler's reward would be the fame and fortune he hoped the project would bring him. Work on the book had possibly started by 1607. One or two boxes of fresh flowers from the garden were being despatched weekly to Nuremberg for Besler to copy. The Prince-Bishop's calculation that the book would cost him 3000 florins turned out to be a very optimistic prediction.

Flower paintings commissioned as a record of favourite or rare specimens emerged as a new genre at the beginning of the seventeenth century. A collection of such paintings was known as a florilegium. This word first appeared in print in Adrian Collaert's *Florilegium*, published in Antwerp in 1590. Pierre Vallet drew and etched 75 plates for *Le Jardin du Roy très Chrestien Henry IV* (1608). A number of them were copied in *Florilegium novum* (1611) by Johann Theodor de Bry. They were not always intended for publication. Johann Walther, an artist of Strasbourg, painted choice blooms at Idstein near Frankfurt-am-Main for Count Johann of Nassau. Camerarius's own florilegium, now in Erlangen University, may have persuaded the Prince-Bishop of Eichstätt to emulate him. Some of the drawings in this florilegium could have been executed by Joachim Jungermann, brother of Ludwig Jungermann who worked with Besler on the *Hortus Eystettensis*. Did Besler ever see this florilegium? Whatever the *Hortus Eystettensis* may have owed to Camerarius, it stands on its own as one of the finest florilegia ever printed. Prince-Bishop Johann Conrad von Gemmingen, for whom the book was an expression of pride in his garden and a celebration of God's bounty, never saw it finished. He died on 7 November 1612, some six months before it was published. At the time of his death 50 plates had been engraved and he had spent 7500 florins on the project. His successor, Prince-Bishop Johann Christoph von Westerstetten, continued his predecessor's building programme and authorized the book's completion.

News about the book's progress had been circulating for some years and, at last, in the summer of 1613 it emerged, a massive royal folio (22½ by 18 inches or 57 by 46 cm). Robert Burton's *Anatomy of melancholy* (1621) extolled it as the 'voluminous and mighty herbal of Besler of Nuremberg, wherein almost every plant is to his own bigness'. It clearly set out to impress, but its bulk may also have been the consequence of Besler's desire to have the plants drawn life-size, as Burton observed. It exudes confidence. A splendid title-page portraying God guiding Adam through the Garden of Eden proclaims its patron's piety. The inclusion of his coat of arms announces his association with it.

The engraved plates of plants, 367 altogether, are grouped under the four seasons: Spring (134), Summer (184), Autumn (42) and Winter (7). With no systematic plant classification available, Besler's arrangement was a reasonable one. Two versions of the book appeared in 1613. In the de luxe edition the plates were printed on superior paper, the verso of each plate being left blank and the accompanying letterpress printed on separate leaves inserted between the plates. This edition was intended for hand-colouring and destined for presentation or for purchase by rich collectors. A trade edition with the text printed on the verso of the plates and, therefore, not intended for colouring, was dedicated to Prince-Bishop Johann Conrad von Gemmingen. Copies were also dedicated to Prince-Bishop Johann Christoph von Westerstetten, who had given or sanctioned funds for finishing the book. Nicolas Barker has cautioned that there is no clear distinction between various printings. It is believed that the combined number of copies printed was 300.

Besler made certain that his own participation in organizing printers, engravers, artists and colourists was not overlooked. His coat of arms and portrait greet the reader at the beginning of the book. This exercise in self-promotion is perpetuated in the four subtitles, whose design is based on an apothecary's equipment (Besler's profession) and, once again, his coat of arms.

Besler acknowledged the advice given by Camerarius and Clusius in his introductory remarks but no other collaborator was thanked: neither Ludwig Jungermann, who helped

Basil Besler *Hortus Eystettensis*, Altdorf 1613. B.L.10.Tab.29.

ABOVE LEFT Plate of tulips, mostly cultivated varieties.
ABOVE RIGHT A spectacular *Lilium bulbiferum* flanked by *Centaurium erythraea*.
The date of colouring of this engraving by Georg Mack appears below the image.

BELOW LEFT The cyclamen has long been cultivated in Europe. J. Gerard's *Herball* (1597)
described two and J. Parkinson's *Paradisi in sole* (1629) ten species and varieties.
BELOW RIGHT The common sunflower (*Helianthus annuus*) arrived in Europe from
Central America and Peru during the sixteenth century.

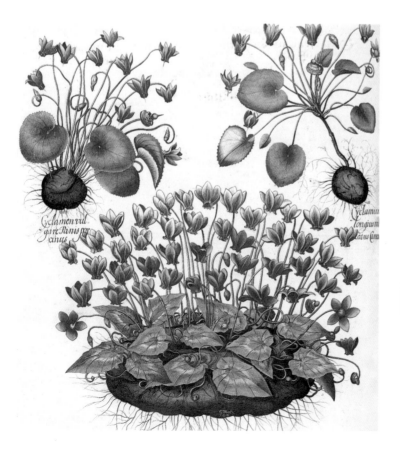

Ficus Indica Eystetten
sis ex uno folio enata lu,
xurians

Basil Besler *Hortus Eystettensis*, Altdorf 1613.
The prickly pear (*Opuntia ficus-indica*) reached Europe from America in 1565. Many of the plants in the *Hortus Eystettensis* are drawn life-size, but since this was not possible with this huge opuntia, a rule was included to give an idea of its bulk. Stadtbibliothek, Nuremberg, Solg.1797a.2°°.

with botanical descriptions and nomenclature, nor his brother Hieronymus, who is believed to have translated the introduction into Latin. Apart from Besler's claim that he himself drew many of the plants there is no mention of any artists. Nicolas Barker proposes a plausible case for including Sebastian Schedel, a Nuremberg artist, who may have made preliminary sketches. Besler almost certainly made use of other florilegia, and possibly that of Camerarius.

A work of the magnitude of the *Hortus Eystettensis* needed the services of a large team of engravers. Wolfgang Kilian, whose skills are demonstrated in the title-page's elaborate design, is the best known. His stepfather, Dominicus Custos, also contributed a few plates. Kilian operated from Augsburg; other engravers had their workshops in Nuremberg. Since relatively few of the plates are signed or carry initials it is not possible to estimate how many were employed, but a dozen would be near the mark. Johann Leypoldt executed the title-page for each of the four seasons as well as floral plates. Several engravers had learned their craft in the Low Countries or were associated with that region. The father of Friedrich and Levin van Hulsen was a Dutchman who had settled in Nuremberg. Peter Isselburgh, also of Nuremberg, may have studied under the celebrated Dutch engraver Crispin de Passe. Other engravers include Heinrich Ulrich, Dietrich Krüger, Hieronymus Lederer, Servatius Raeven and G. Remus.

These engravers needed muscular dexterity to force the sharp wedge of the burin across the surface of large copper plates, at the same time controlling the depth of penetration which determined the width of lines. Tonal values were created by conventional cross-hatching. The pervasive coarseness of these engravings is concealed by pigment in coloured copies of the book.

As with the engravers, there is no certainty about the number of colourists employed. Some are indicated by enigmatic initials on the plates, but one colourist was not constrained by such modesty. Georg Mack the Younger, who ostentatiously wrote his name in full on the title-page of the copy in The British Library, came from a Nuremberg family of colourists or illuminators. The time taken to colour this particular copy is revealed by the dates written on the plates by Mack: 2 March 1614 on the title-page and 16 April 1615 on the last plate. He probably used the original paintings as a colour guide. Georg Schneider coloured the title-pages of the first two copies to be finished. Magdalena Fürstin, who had received painting lessons from M.S. Merian, coloured plates in the copy now in the National Library in Vienna.

When several colourists are employed, colour variations are likely to occur. In a comparative study of coloured copies of the *Hortus Eystettensis*, Barker has discovered that colour tints for the same flower often vary and on occasions are entirely different.

Such errors or indifference to accuracy are not surprising in a book where beauty and ostentation are paramount. Leaves are contorted to integrate in an overall decorative pattern. Symmetry and balance are guiding principles in the disposition of plants usually arranged in trios and quartets on the page. This numerical presentation, mechanically repeated, ultimately robs the design of any spontaneity. Plate after plate conforms to a rhythmic composition of stems, leaves and roots. Some flowers are choreographed to dance on the page: the three *Muscari*, for instance, linked by their leaves, or the chorus of tulips, are vaguely reminiscent of images in Walt Disney's *Fantasia*. Family or generic relationships are of little consequence: *Picea* cones partner two twigs of cherry blossom, and four snippets of *Vinca minor* escort a *Staphylea*. A balanced design was all that mattered to Besler, who presumably ultimately decided what went on each page. Many plants are depicted life-size. Where they were too tall for the page a section of the stem was removed; the inflorescence of *Lilium martagon* appears on one plate and its bulb on another. These comments are not intended to diminish the artistic merits of a book which is a landmark in floral illustration. It is a joy to look at, but in order to appreciate its astonishing vitality only a few plates at a time should be studied.

The garden at Eichstätt was one of the best stocked in Germany and the *Hortus Eystettensis* is an invaluable record of what was being grown in some European gardens during the late sixteenth and early seventeenth centuries. Its figures of over 1000 species and cultivated varieties are, unfortunately, not always identifiable with any certainty.

Southern Europe, the Mediterranean region and the Middle East are represented by more than 200 species. Medicinal plants include the bay tree (*Laurus nobilis*), the sea onion (*Urginea maritima*) and *Aristolochia rotunda*, known in England as birthwort from its use in childbirth. Fashionable plants obviously got preferential treatment. Narcissi found mainly around the Mediterranean have 41 figures of which seven are Tazettas.

Many of the 22 hyacinths, native to the Balkans, Greece and Asia Minor, are *Hyacinthus orientalis*. The Padua botanic garden first introduced it into cultivation. Besler included twelve drawings of the gillyflower (*Matthiola incana*), commonly found in European gardens from the mid-sixteenth century. Turkey's great contribution was the tulip. Hainhofer stated that Eichstätt had about 500 varieties, a questionable claim, but there are more tulips than any other flower in the *Hortus Eystettensis* – 54, mostly garden varieties. Three stunning plates are allocated to the flower head of the crown imperial (*Fritillaria imperialis*). A native of Persia, Afghanistan and the Himalayas, this plant was introduced to Vienna along with the tulip by Clusius, to whom gardeners were also indebted for new anemones, hyacinths, irises, lilies and ranunculus.

Plant introduction from the New World was still spasmodic, but the *Opuntia* and *Agave* given prominence on the title-page of the *Hortus Eystettensis* hinted at the promise of a spectacular flora. Most of the 20 or so plants featured by Besler came via Spain from Central and South America. He gave two plates to the tomato, then considered a potent aphrodisiac. Grown for its attractive flowers, it would be some years before the potato became a staple food. Tobacco is believed to have been smoked in Germany from the early seventeenth century. Chilli pepper (*Capsicum* sp.), originally admired as an ornamental plant, was depicted in fifteen drawings. Another favourite was *Mirabilis jalapa*, known as the Marvel of Peru. With its spectrum of colours Gerard's *Herball* (1597) rated it 'rather the Marvel of the World, than of Peru alone'. *Tagetes erecta*, the African or French marigold, was one of a number of showy plants from Mexico. The tall sunflower (*Helianthus annuus*), revered by the Incas of Peru as a symbol of their sun god, astonished European gardeners, who competed to grow the biggest specimens. Camerarius said it was to be found in every garden. Besler's portrait of its flower head is a bold and striking portrait. North America is poorly represented, with only the white cedar (*Thuja occidentalis*), one of the few trees in the book, and *Anaphalis margaritacea*.

About half the plants in the *Hortus Eystettensis* had medicinal properties attributed to them, just under 200 were used in cooking and the rest were valued as ornamentals.

The *Hortus Eystettensis* was sold in loose leaves without any pagination, making its assembly in correct order a challenge for bookbinders. The plain edition with the text printed on the verso of the engraved plates sold for about 35 florins. Coloured copies cost 500 florins, a price queried by one purchaser who assumed that 50 florins had been intended! Besler most likely received the proceeds from the sale of the plain version, which sold well enough for him to be able to move into a fine house in Nuremberg. At the outset sales appear to have been confined to Germany. After the initial print run of about 300 copies he enjoyed free use of the plates for four years. By 1617 the price of a plain copy had risen to 48 florins. Barker has discovered several copies dated 1627 with just a sampling of the engravings, presumably made up from spare sheets.

Interest in the book did not subside with Besler's death in 1629. Prince-Bishop Marquard II reprinted the plates without the text in 1640, hoping it would sell in France, the Low Countries and England. Besler's name was removed from the title-page although his family still retained an interest in the book. Under the patronage of Prince-Bishop Johann Anton I a centenary edition was planned, with the inclusion of some additional plates. Even a new title-page with the date '1713' was engraved, but it was not until 1746 with the help of the Prince-Bishop's personal physician, C.J. Trew, that it was finished and published (still with the '1713' date) in about 1750. Besler's name was not reinstated. Sales were disappointing – maybe because the book-buying public wanted illustrations of the latest exotic plants; in 1817, 86 copies still had not been sold.

A German bibliographer, Claus Nissen, recorded that in December 1820 the original copper plates had been seen in a library in Augsburg. For many years it has been believed that some time after that date they were melted down by the Munich mint, but in 1998 329 plates, including that of the splendid title-page, were discovered in the Albertina Graphic Collection in Vienna.

Much of the original garden at Eichstätt disappeared under the foundation of fortifications erected during and after the Thirty Years' War. The 'centenary' edition of the *Hortus Eystettensis* referred to the lost garden as *olim* (once, or formerly). In 1998 the Bastion Garden was created on part of the site, stocked with plants depicted in the book and displayed in flower beds grouped appropriately according to the seasons.

Although the original garden has gone, the book remains its memorial and also a reminder of generous patrons, self-effacing botanists and of a team of craftsmen directed by the indefatigable Besler.

Portrait of Basil Besler holding a sprig of basil in his hand, by Lorenz Strauch.
Engraved by W. Kilian. Besler's coat of arms is in the accompanying oval. B.L.10.Tab.29.

5 Some early nurserymen's catalogues

EMANUEL SWEERTS (1552–1612)

Some of the flowers grown in the palace garden at Eichstätt were supplied by Emanuel Sweerts, a Dutch merchant. Born in Zevenbergen in northern Brabant, he worked in Amsterdam for much of his life, both as an artist and as a dealer in curios, natural history objects, bulbs and seeds. He was well known to naturalists and gardeners and to the Hapsburg Emperor, Rudolph II, a wealthy collector and patron of the arts, and his most prestigious client. An offer of the post of Prefectus or Director of the royal gardens in Prague was declined by Sweerts, who nevertheless was pleased to accept an invitation to compile an album of the Emperor's rare and interesting flowers for engraving and publication under royal patronage. Although Rudolph purchased the set of floral illustrations in 1609, several years elapsed before they went to press. Rudolph died in 1612 and the work was consequently dedicated to his brother, Matthias.

Tersely entitled *Florilegium*, it was published in Frankfurt in 1612 to coincide with the annual trade fair, which Sweerts visited every year. An advertisement on the verso of the title-page announced that the book and the plants it illustrated could be purchased at the author's shop near the Frankfurt Römer for the duration of the fair and thereafter at the premises of Paulus Aertssen van Ravestyn, a printer in the Bloemengracht in Amsterdam. No prices are quoted, but it is without doubt a nurseryman's catalogue, probably the first of the genre. Sweerts is portrayed within an oval frame on whose rim is inscribed the legend 'Vita hominum flos est' (The life of man is like that of a flower). His right hand lightly resting on a human skull reaffirms the message of man's transience. The title-page, a composition of classical figures in front of a formal garden, is heightened with gold in some coloured copies.

It is a reasonable supposition that Sweerts drew some of the plants himself, but there is no confirmatory evidence of this. A number of the plant portraits, with some modifications, are similar to those in Johann Theodor de Bry's *Florilegium novum* (proof copy in 1611), but who copied whom is a matter for conjecture. Apart from the author's preface, there is no text. Part 1 has 67 engraved plates of bulbous and tuberous flowers, and part 2 has 43 plates of those with fibrous roots ('fibrosae radices') – altogether 110 plates. More than 560 flowers are depicted, usually life-size. The crown imperial, a favourite with most gardeners, leads the way. Thirty-three tulip heads are paraded in regular columns. An abundance of irises, lilies and narcissi confirms their popularity. There would be eager customers for the varietal forms of anemones, crocuses, fritillaries, snowdrops, auriculas, campanulas, poppies and cyclamen. The American contingent is represented by the sunflower, cactus, agave, pineapple and Canna.

Emanuel Sweerts *Florilegium*, Frankfurt 1612. Title-page.
The figures represent Apollo the sun god, Flora (seated) and Diana the moon goddess,
posed before a formal garden. The opuntia and agave are prominently depicted, as they are on the title-page
of the *Hortus Eystettensis*. Above is the Tetragrammaton or the Hebrew name for God. At the bottom
are the portraits of Clusius and Dodoens, two leading botanists. B.L.451.i.5.

Sweerts died in 1612 before he could enjoy the financial benefit of his catalogue's popularity as a florilegium. There were further reprints in 1614, 1620, 1631, 1641, 1647 and 1655, notwithstanding some deterioration in the quality of its plates.

CATALOGUS PLANTARUM (1730)

English gardeners had bought, sold and probably exchanged plants and seeds long before an organized trade emerged – at least in London – during the Tudor period. Nurseries, as we understand the term today, made their debut about the middle of the seventeenth century. A pioneer, Leonard Gurle, converted some land between Spitalfields and Whitechapel into one. One of the earliest surviving nurserymen's catalogues, with prices of its produce, was issued in 1688 by George Ricketts of Hoxton, then a flourishing centre of the trade. The most important nursery at that time was formed in 1681 at Brompton Park (part of the site is now occupied by the Victoria and Albert Museum) by four former head gardeners. By 1705 it had millions of plants under cultivation on a site of well over 50 acres. Thomas Fairchild at Hoxton, Robert Furber at Kensington and Christopher Gray at Fulham emulated its success. Some nurserymen produced gardening manuals as a means of enhancing their reputation and of publicizing their business. Thomas Fairchild brought out *The city gardener* in 1722 and John Cowell three editions of *The curious and profitable gardener* between 1730 and 1733.

The *Catalogus plantarum* (1730) was sponsored by the Society of Gardeners, a trade association mainly of nurserymen and seedsmen in London, formed in the early 1720s. About 20 in number, its members each brought specimens of their stock to monthly meetings at Newhall's Coffee-house in Chelsea for discussion and entry in a register of approved names. They aimed to standardize plant names, thereby resolving the confusion in nomenclature which frequently led to the same plant being sold under different names, much to the annoyance of their customers. As well as being documented, a

Emanuel Sweerts *Florilegium*, Frankfurt 1612. B.L.451.i.5.
LEFT Plate 8. Flower heads of twelve tulips.
RIGHT Plate 39. *Iris oncocyclus*, based on plate 60 in Theodor de Bry's *Florilegium novum* (1611). Sweerts also took illustrations from Matthiolus, L'Obel and Clusius.

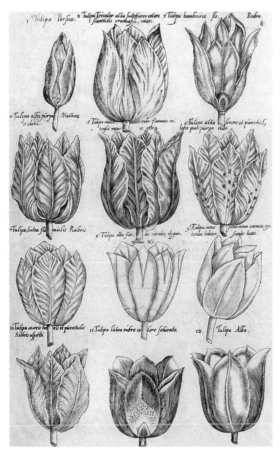 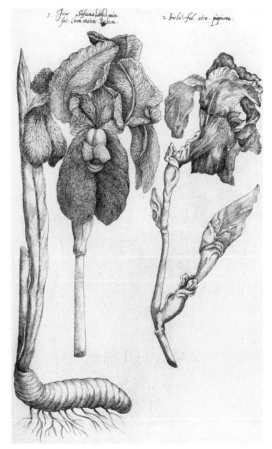

Society of Gardeners *Catalogus plantarum*, London 1730. B.L.452.h.2.
LEFT Frontispiece. A garden vista of clipped yew hedges terminated by a distant conservatory, partially framed by flowers and fruit in the foreground. Engraved by Henry Fletcher in 1729.
RIGHT Plate 16. Two *Pinus* species by Jacob van Huysum, engraved and printed by Elisha Kirkall in a combination of etching and mezzotint. Colour-printed, *à la poupée*, in three colours.

selection of the plants was drawn for reference purposes. Eventually they had accumulated sufficient data to merit publication in the form of a list of English and corresponding Latin names. Their clients persuaded them to include cultivation notes as well. With the addition of engravings of some of their flower paintings, the original concept of a list evolved into a handsome book. The selection of plants was restricted to those available in the nurseries of the Society's members, who undertook to offer customers advice and, presumably, to fulfil orders.

Dedicated to Lord Pembroke, the owner of renowned gardens at Wilton House, the *Catalogus plantarum* was published anonymously in February 1730. The putative author, Philip Miller, was secretary of the Society of Gardeners and head gardener at the Chelsea Physic Garden. This slim folio on hardy trees and shrubs was to be followed by further volumes on greenhouse and stove plants, ornamental flowers, fruit trees and shrubs. Only this first part appeared, one of the earliest flower books to have some of its plates not coloured by hand but printed in colour.

In an attempt to eliminate colour variability, which marred hand-coloured engravings, experiments were made with the application of printers' inks (which usually retained their colour uniformity) to individual copper plates using the *à la poupée* method. Elisha Kirkall tried the process in John Martyn's *Historia plantarum rariorum* (1728–37),

The key at the bottom of the plate reads:

1 Duke Vantol Tulip.
2 Silver Edg.d Alaternus.
3 Yellow bleach Alaternus.
4 Cornelian Cherry.
5 White Mezereon.
6 Red Mezereon.
7 Double Narcissus. of Constantinople.
8 Single Anemone Purple & White

9 Venetian Vetch—true
10 Double blew Hepatica.
11 Early white Hyacinth.
12 Blush red Dens Caninus.
13 Spring Cyclamen white Edg.d
14 Strip'd & Edg.d Polyanthos.
15 Single white Hepatica.
16 Single blew Hepatica.
17 White Dens Caninus.

FEBRUARY

18 Double Peach colour'd Hepatica
19 The greater Snow-drop.
20 White Crocus.
21 Double Snow-drop.
22 Small yellow Crocus.
23 Great blew Crocus.
24 Small blew Crocus.
25 Single dark-red Anemone.
26 Pantaloon Strip'd Polyanthos.

27 Persian Iris.
28 Yellow dutch Crocus.
29 Scotch white strip'd Crocus.
30 Blew Hyacinth Passe tout.
31 Fruit bearing Almond.
32 Single Prussian blew Anemone
33 Yellow Colutea.
34 Peach colour'd single Hepatica
35 Double Pilewort.

Robert Furber *Twelve months of flowers*, London 1730–32. Plate for February.
Drawn by Pieter Casteels and engraved by Henry Fletcher. Each monthly bouquet displays about 33 or 34
flowers, all numbered, with a key at the bottom of each plate which is inscribed 'From the collection
of Robert Furber, Gardener at Kensington'. B.L.10.Tab.45.

combining mezzotinting with etching; tonal values were added to the etched lines on a copper plate with a mezzotint rocker. As the range of ink pigments at his disposal was limited, Kirkall seldom used more than three of them to ink appropriate parts of the plate with green and brown as basic colours. The printed impression was subsequently touched up by hand.

He used the same process on seven of the 21 plates in the *Catalogus plantarum*, the remainder being etchings by Henry Fletcher and coloured by hand. The flower paintings for both this book and Martyn's were executed by Jacob van Huysum (*c.*1687–1740), the less successful brother of the Dutch flower painter Jan van Huysum. He had come to London in 1721 and was employed by Sir Robert Walpole to make copies of old masters. He drank to excess, providing Horace Walpole with numerous anecdotes about his behaviour to amuse his correspondents. In neither book are the plates outstanding.

More than 80 trees and shrubs from North America are listed in the *Catalogus plantarum*, a bias that must have delighted two of the Society's members, Robert Furber and Christopher Gray, whose nurseries specialized in the flora of that part of the world. It was Gray who was to publish Catesby's *Hortus Britanno-Americanus* in 1763. The *Catalogus plantarum* included ten trees and shrubs raised from Catesby's seeds. The popularity of American plants can in part be attributed to the relatively short North Atlantic crossing, which increased the survival rate of plants in transit. Furthermore, many of the plants from the eastern seaboard were hardy in England. Rather out of context in this assemblage of trees are a plate of roses and a splendid 'white lilly striped with purple'. The British Museum has a collection of 141 drawings of flowers and fruits by Van Huysum originally destined for use in future parts of this aborted publication.

ROBERT FURBER (*c.*1674–1756)

Of all the publications produced by members of the Society of Gardeners, Robert Furber's displayed the most originality. He founded his nursery in the Kensington Road west of Hyde Park Gate early in the eighteenth century. His business prospered as his stock grew, much of it imported, and in 1724 Philip Miller paid him a compliment by listing a selection of his exotic trees and shrubs in his *Gardeners and florists dictionary*. In 1727 Furber issued his own catalogues, one on fruit varieties and the other on English and foreign trees. These were modest pamphlets, but his next catalogue, *Twelve months of flowers*, was extravagantly illustrated in the grand manner.

However, its illustrations did not constitute a conventional portrayal of plants such as that, for instance, in the *Catalogus plantarum*. Furber engaged Pieter Casteels (1684–1749), a Flemish artist resident in London since about 1708 and much in demand by the gentry to paint flowerpieces to hang above their fireplaces and doors. Casteels executed twelve compositions of flowers bunched in ornamental vases and urns, done in the Flemish and Dutch styles, one for each month of the year.[22] Each presented a baroque bouquet of

---

22. These paintings are now at Fort Worth, Texas.

more than 30 different flowers, all from Furber's nursery. Casteels had completed his commission by 25 September 1731, when *Fog's Weekly Journal* reminded subscribers to this novel catalogue to send the balance of their subscriptions to Furber, Casteels or Henry Fletcher, who engraved the paintings. A thirteenth plate with subscribers' names within a floral border appeared in March 1732. It was dedicated to Frederick, Prince of Wales, the Princess Royal and all the 435 subscribers, among whom were peers and gentry and notable garden enthusiasts such as Peter Collinson, J. J. Dillenius, Lord Petre, Isaac Rand and James Sherard (who had subscribed for three copies). Sets of the twelve plates were available to non-subscribers at one pound five shillings (plain copy) or two pounds twelve shillings and sixpence (coloured copy).

A number against every flower, which customers quoted when ordering their selection, identified it in a key at the bottom of each plate. Obviously, some of the flowers never bloomed in the months in which they appear, especially those for autumn and winter – a liberty that Dutch flower painters often took with their compositions. Lord Orford, who was not a devotee of this genre of painting, said of Casteel's work that they 'have neither the boldness and relievo of a master, nor the finished accuracy that in so many Flemish painters almost atones for the want of genius'.

Furber's selection was calculated to appeal to florists, that is, gardeners who specialized in cultivated varieties of certain decorative plants and who often exhibited their choice blooms. Their favourite was the auricula, of which there were more than 200 varieties in cultivation during the 1730s. Furber included 26 of them and 19 of the anemone, probably the next in popularity. Other florists' flowers such as the hyacinth, tulip and ranunculus are well represented.

The instant success of this innovative catalogue was quickly followed by a pirated version which was denounced by Furber, Casteels and Fletcher in *Fog's Weekly Journal* for 18 March 1732:

> *Whereas there is advertised in the Universal Spectator and other papers, a New Edition of the most curious and uncommon Flowers of the Twelve Months in the Year, Notice is hereby given to all Noblemen, Ladies and Gentlemen, that those Prints are only Copies, and that we are not any ways concern'd in 'em, but the Originals, which are only sold by us, Robert Furber, Gardener at Kensington, Peter Casteels, Painter in Long Acre, Henry Fletcher Engraver in Nottingham-street, the upper End of Plumbtree Street, Bloomsbury; and at Mr Tho. Bowles in St Paul's Church Yard, Print seller …*

> *N.B. There's a Thirteenth Plate, in which is a list of the Subscribers Names with a Border of Flowers about it, and our Names engrav'd at the Bottom of Each Plate, and those are the Originals; and all without our Names are Copies. Theirs are not above half the bigness.*

Furber is acknowledged in *The flower-garden display'd* (1732), Casteels's paintings being

Robert Furber *Twelve months of flowers*, London 1730–32. Plate for June.
Drawn by Pieter Casteels and engraved by Henry Fletcher. B.L.10.Tab.45.

re-engraved by James Smith. The book shows all the signs of hurry: inferior engraving and execrable hand-colouring. As an inducement to potential buyers it had an anonymous text (attributed to Richard Bradley, Professor of Botany at Cambridge) on flower cultivation. The scope of the book is extended by suggesting its usefulness for 'Painters, Carvers, Japanners, etc., also, for the Ladies, as Patterns for Working and Painting in Water-Colours, or Furniture for the Closet'. A second edition in 1734 included an account on 'the art of raising Flowers in the depth of winter, in a Bed-Chamber, Closet, or Dining-Room. Also the method of raising Salleting, Cucumbers, Melons, etc at any time in the year.'

The book spawned a progeny of variant versions during the 1740s. Print dealers offered copies of degenerate images which had been reversed during the copying process. The prints were made more decorative by adding butterflies and other insects. Robert Sayer, John King and Phil Overton sold such a set. John Bowler published another set of

Robert Furber *Twelve months of fruit*, London 1732. Plate for June. The success of *Twelve months of flowers* encouraged Furber to offer a selection of fruit in a similar format, again with Casteels and Fletcher as artist and engraver. The British Library copy is plain, but coloured copies were also available. B.L.45b.h.23.

the plates in 1749 under the title of *Flora, or a curious collection of ye most beautiful flowers as they appear in their greatest perfection each month of the year*. The book's influence extended beyond the book and print trade. In the Toronto Museum, for example, there is a coverlet which incorporates Casteels's paintings as floral motifs.

Anticipating another winner, Furber quickly followed *Twelve months of flowers* with *Twelve months of fruit* (1732). The formula was the same: plates by Casteels, one for each month, the fruits selected from Furber's nursery and all keyed by numbers to their names on the lower edge of the prints. An extravagant sampling of the produce of English orchards is piled into bowls, baskets, plates or on a plain table top: 364 varieties of apples, pears, peaches and soft fruit (currants, gooseberries, raspberries, strawberries). Five engravers, including Henry Fletcher and James Smith, copied this abundant harvest. Their evocative names in Furber's two books poignantly remind us of the many varieties we have lost.

6 Professor Beringer's fossils

About ten years ago an antiquarian bookseller's catalogue included an exceedingly rare item – the 1726 edition of Beringer's *Lithographiae Wirceburgensis*. An annotation explained that it was 'the result of a famous hoax' perpetrated by some of Beringer's students. They had carved small stones with bizarre designs, burying them with the intention that he would find them during his excavations. With incredible gullibility, he accepted them as genuine and published this book with engravings of them. It was only when he found another 'fossil' with his name on it that he realized he had been duped. Desperately he tried to destroy all existing copies, buying back those that had been sold. This distressing story of deception and humiliation is a well-known episode in the history of palaeontology. It is partly a legend, however, and one that has been embellished through retelling, but through the writings of M.E. Jahn, D.J. Woolf and others we now know the true facts. In order to understand the naivety of this unfortunate academic we need to put the incident in the context of scientific knowledge at that time.

Eratosthenes, Strabo, Xanthus, Xenophanes and other Greek philosophers had reasoned that fossil shells once contained living creatures, gradually over aeons transmuting into stone and providing evidence that dry land had once been covered by the sea. During the Middle Ages the credulous believed that large fossil bones were the remains of giants and dragons. Others viewed fossils as inorganic substances, fortuitously shaped, rather than petrifactions. Christians inclined to the theory that fossils had been creatures drowned in the biblical Great Flood. By the seventeenth century there were a number of propositions from which to choose. The British naturalists Edward Lhwyd and Martin Lister accepted that fossils were probably natural freaks, not vestiges of former organisms. Robert Plot had a predilection for *vis plastica* or 'plastic virtue' in the ground shaping them. John Ray, undecided, had reservations about *vis plastica* and accidentally formed stones, and rejected the Great Flood hypothesis. Lhwyd corresponded with Johann Scheuchzer, a Swiss doctor, the author of the first illustrated account of fossil plants. In his *Herbarium diluvianum* (1709) Scheuchzer abandoned an earlier belief in figured stones, preferring the Great Deluge explanation. Now committed to diluvialism, he identified some fragments of fossilized vertebrae as relics of a wretched victim of the biblical flood. These he solemnly named *Homo diluvii testis*. Nearly a century later a French palaeontologist pronounced Scheuchzer's fossil to be that of a giant salamander, renaming it *Andrias scheuchzeri*.

Johann Bartholomaeus Adam Beringer (*c.*1667–1738), Senior Professor and Dean of the Faculty of Medicine at the University of Würzburg, was well known for his researches in oryctics, the identification of objects dug up out of the ground. He had built up a cabinet of natural rarities collected in Germany and neighbouring countries,

Johann Bartholomaeus Adam Beringer *Lithographiae Wirceburgensis*, Würzburg 1726. Frontispiece. A mound of 'fossils', perhaps meant to symbolize Mount Eivelstadt, where the notorious finds were made. B.L. 443.h.15.

but the locality that yielded him most specimens was a barren hill at Eivelstadt, a mile or so outside Würzburg. In May 1725 he engaged three youths – Christian Zänger and two brothers, Niklaus and Valentin Hehn – to excavate it. Over the next six months his helpers dug up fossil mollusca and hundreds of figured fossils, all seldom more than three inches long. An early find featured a radiant sun with a human face. Other heavenly bodies followed: a crescent moon, stars, and comets with tails. The limestone fragments carried reliefs of lizards, fishes, bees clinging to flowers, spiders still on guard in their webs, even some copulating frogs. The greatest find, filling Beringer with reverential awe, was a piece with Jehovah's name in Hebrew. He invited some of his university colleagues to witness these astonishing excavations and to share his triumph.

But not everyone congratulated Beringer. In his book he mentioned two academic colleagues (whom he did not name) who publicly denounced the fossils as fakes. J. Ignatz Roderick, Professor of Geography, Algebra and Analysis, and Georg von Eckhart, Privy Councillor and Librarian, both at the University of Würzburg, were the two dissenters. They spread a rumour that Beringer's diggers had fabricated these fossils, then pretended to have dug them up. Perhaps they tried to discredit Beringer when they learned with some alarm that he intended writing a book on his discoveries. While Beringer was prepared to concede that some fragments might not be genuine, he could not believe that any forger would go to the trouble of making so many.

He wrote his book in the evenings after his official duties, evidently in some haste since it was published in 1726. *Lithographiae Wirceburgensis* portrays more than 200 of these disputed fossils on 21 plates. It was presented as the doctoral dissertation of Georg Ludwig Hueber, one of Beringer's students. Universities at that time required candidates to defend a thesis proposed by their supervisor and, moreover, to pay for its publication. As one reads the admirable commentary and translation by Messrs Jahn and Woolf, one senses Beringer's confidence and doubts, his defiance and anxiety.

He examined his unusual specimens in the context of current conflicting theories on fossils. He admitted that the images on the stones, so sharply defined and smoothly polished once the soil on them had been removed, suggested a sculptor's hand. Here and there marks and gouges on the surface looked suspiciously like the work of a careless knife. He refuted the proposition that they were pagan relics. Such theories as the 'plastic force of light', 'vapours rising from the ocean', rocks impregnated with seeds or spermatozoa and the Great Flood were dismissed. Divine intervention might have imposed the Hebrew script on ten of the stones. Why not, he wrote, acknowledge the 'Author of Nature' as the creator of these strange stones, a solution that was 'one of piety and expediency rather than of erudition and the science of physiology'.

He looked dispassionately at the evidence for a hoax: the unnatural distribution of the fossils on the hill itself, for instance. He ridiculed the suggestions that his three youths had carved them, since they patently lacked the skill and the knowledge to reproduce the forms of exotic plants and animals. Many of the stones had been dug out of scrub where there was no trace of soil disturbance, which would have been visible had they been previously buried there. Nor would it have been easy for his diggers to 'plant' them unobserved.

Beringer expresses bewilderment at the behaviour of his two antagonists: secretly plotting against him, spreading malicious rumours, never willing to enter into any public debate. Furthermore, they were inconsistent in their accusations, at one time declaring that potters had fabricated the fossils with baked clay, then deciding they had been carved with a knife. He could only assume that they were motivated by jealousy. With his professional reputation and personal integrity at stake, Beringer requested that the three men in his employ be interrogated by the Chapter of Würzburg Cathedral to establish whether they were guilty of this deception.

In 1934 a German scholar, Heinrich Kirchner, unearthed in the state archives a transcript of the judicial proceedings held in 1726. His paper on this discovery, published in 1935, was forgotten until Carl C. Beringer resurrected it in his *Geschichte der Geologie und des geologischen Weltbildes* (1954). *The lying stones* (1963) by M.E. Jahn and D.J. Woolf includes the 1726 inquiry which destroyed much of the fiction surrounding Beringer.

The first stage of the inquiry, conducted in Würzburg Cathedral on 13 April 1726, set out to establish whether the three youths had the ability to carve stones, to ascertain whether they had seen any of the figures in any book, and to discover how they found the stones or whether they had seen anyone burying them on the hill. The most pertinent question related to Roderick and Eckhart. Had the youths been persuaded by these two men to carve the stones and subsequently to pretend they had been dug up? Roderick and Eckhart were clearly under suspicion and the investigators wanted to question any fellow conspirators.

On 15 April 1726 the proceedings continued in the town hall at Eivelstadt. Niklaus Hehn, aged 18, was the first of the trio to be questioned. He admitted finding the stones on the hill but denied carving or polishing any of them. Roderick had made threats against his younger brother if he did not confess to carving the stones. Zänger had told him that Roderick had carved a sea horse on a stone. Valentin Hehn, aged 14, confirmed his brother's statement.

Johann Bartholomaeus Adam Beringer *Lithographiae Wirceburgensis*, Würzburg 1726. B.L.443.h.15.
LEFT Sun, moon and comets. CENTRE Collection of birds, two with eggs and a skeleton of another.
RIGHT Fragments of scripts, some possibly Hebrew.

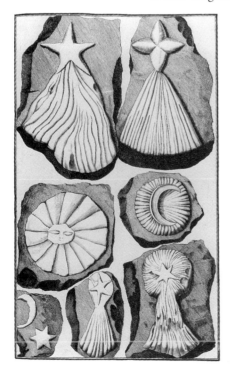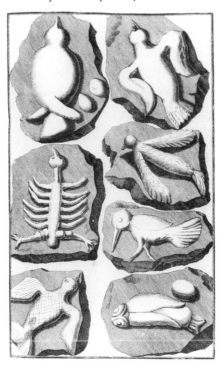

Two of J.B.A. Beringer's 'fossils', one with some lettering, the other with a slug-like creature.
University Museum, Oxford.

The testimony of Christian Zänger, aged 17, further implicated Roderick when he accused him of carving the stones and paying him to take some of them to Dr Beringer: one had a dragon on it, another a pomegranate and a third a lion. Roderick had asked him to obtain some pieces of hard marble and to polish them. Roderick had shown him animal drawings and Hebrew script, which Zänger claimed were featured on stones still concealed on the hill. Most damning was his recollection of a conversation he had overheard between Roderick and a Baron von Hof, in which they discussed a plan to humiliate Dr Beringer because 'he was so arrogant and despised them all'.

The proceedings were adjourned until 11 June, when the examination of Zänger, now the key witness, continued. He confessed to polishing stones in Eckhart's house. He said that Dr Beringer had declined to buy some of the stones offered to him when he suspected they were not genuine.

At this point the archival record suddenly and inconclusively ends. We do not know the fate of the three boys; Eckhart was denied access to the city archives, essential for his research, and died four years later; Roderick left Würzburg, probably banished. In 1730 he petitioned the Prince-Bishop of Würzburg to allow him to return in order to write a biography of his late friend Eckhart, a request grudgingly granted.

Beringer was vindicated. He returned to academic life, wrote a couple more books and died fourteen years later. How did some of his near contemporaries judge him? Johann Gesner's *Tractatus physicus de petrificatus* (1758) saw him as a 'malicious imposter' who made 'opprobrious sport of the science of lithology'. James Parkinson in his *Organic remains of a former world* (1804) pointed to his impetuosity as a salutary lesson to be 'more cautious in indulging in unsupported hypotheses'.

A so-called second edition of *Lithographiae Wirceburgensis* was published in 1767, almost 30 years after Beringer's death. His name appears at the top of the title-page while that of his former student Hueber is omitted. According to Jahn its text pages were most likely retrieved from copies of the original edition recalled from booksellers. Today's rarity of the 1726 edition is due not to its deliberate destruction by a frantic author, but possibly to its suppression during the trial at Würzburg.

There is a postscript to this story. A Jesuit priest, Athanasius Kircher (1602–80), who coincidentally began his teaching career at Würzburg before finding appointments in France and Italy, was also a fossil collector. He subscribed to a proposition that unusual stones had been formed long ago by a petrifying fluid percolating a volatile Earth. He, too, fell a victim to hoaxers. J.B. Mencke in *De charlatoneria eruditorum declamationes duae* (1715) told of some youths carving fantastic figures on a stone, which they buried on a site where a house was about to be built for Kircher. He found it, as he was meant to, and provided an ingenious interpretation of it. Beringer had discussed Kircher's theories in his book and it is conceivable that Roderick and the learned Eckhart were aware of anecdotes about his credulity. Did Kircher give them the idea to play a similar trick on Beringer?

There have always been forgers working either for profit or pleasure. In the eighteenth century nimble Japanese fingers fashioned mermaids and mermen out of the head and torso of a small monkey and the tail of a fish to sell to credulous Westerners. A seven-headed hydra, once the awesome relic of a Bohemian church, was recognized by Linnaeus as a neat conjunction of the jaws and claws of weasels covered with snake-skins. When the botanist Jussieu and his students fabricated a flower from disparate parts of several plants and showed it to the great naturalist, Linnaeus was not deceived. 'Only Jussieu or God could do so,' he wryly observed.

The deception imposed on C.S. Rafinesque by J.J. Audubon (the subject of a later chapter in this book) comes into a similar category of a practical joke, although a somewhat cruel one. Rafinesque was an eccentric polymath of whom it was once said that 'he missed greatness by embracing too many fields of knowledge'. He lived a peripatetic existence in Italy and France before settling in the United States. While Audubon's guest he smashed his host's favourite fiddle by using it to kill bats invading his bedroom. Audubon got his revenge some time later by presenting him with a set of drawings of a dozen fish with notes on their habitats. They were, in fact, fictitious fish, but the ingenuous Rafinesque declared them to be new species and published them as such in his *Ichthyologia Ohiensis*. Such uncritical acceptance can be ridiculed, but a determination not to be fooled can sometimes rebound on the cautious. When some British zoologists first saw the body of the extraordinary duck-billed platypus, they were certain that a fraudster had been at work.

Faking is still very much alive and flourishing, as the Professor of Zoology at Harvard, Stephen Jay Gould, found out on a recent visit to Morocco. There he was accosted by small boys hawking fossils and saw shops filled with them. Common ones such as ammonites were 'improved' by judicious carving, but others reminded him of Beringer's specimens and he wondered if they were deliberate copies of them. But on closer examination he decided they were 'merely ludicrous and preposterous', not nearly as good as those that had deceived Beringer.

Plate *498*.

Elizabeth Blackwell *A curious herbal*, London 1737–39. Vol. 2, plate 498.
White waterlily (*Nymphaea alba*). Roots and flowers 'stop all kinds of fluxes and gonorrhaea and
nocturnal pollutions by their softening cooling qualities, allaying the acrimony of the seed;
and thereby rendering persons less inclined to venery'. B.L.452.f.2.

7 A curious tale

Mrs Elizabeth Blackwell is remembered today as the author of a single flower book and as the loyal wife of a foolhardy and ill-fated husband. Little is known about her early years. She is believed to have been one of the daughters of a Mr Blachrie, an Aberdeen stocking merchant. One authority asserts that she secretly married Alexander Blackwell (her cousin, according to *The Scotsman*, 11 November 1961) and the couple eloped to London. This is contradicted in *The Bath Journal* of 14 September 1747, by a correspondent who signed himself 'G.J.', in a long article occasioned by the death of Alexander Blackwell. It claims to include 'several particulars not before published relating to the doctor's origins, his conduct and behaviour'. Since these anonymous reminiscences of Blackwell have the ring of personal knowledge, I shall refer to them again. Whatever the circumstances of their meeting and marriage, there is no denying that consequently Elizabeth's life completely changed.

There are contradictions, too, in Alexander Blackwell's life. He was born in 1709, the son of Thomas Blackwell, Professor of Divinity at Marischal College, Aberdeen and later Principal. With parental encouragement he acquired a good grasp of Latin and Greek before he was fifteen. He continued his studies in the classics at Marischal College, adding French to his repertoire of languages. We are told that, now confident that he was well prepared for a business career, he refused to study for a degree with a view to entering one of the learned professions. He left Marischal College and *The Bath Journal* stated that 'he went away so privately that his friends knew not what was become of him, until his arrival in London'. John Nichols wrote that he went to Leiden to study medicine,[23] but R.W.I. Smith's *English-speaking students of medicine at the University of Leyden* (1932) does not record his name. At some stage, however, Blackwell gained enough medical expertise to be able to treat the Swedish king. Nichols adds that on his way back to England he stayed in The Hague, where a 'Swedish nobleman' befriended him – his first contact, and perhaps a significant one, with a native of Sweden.

In London he found employment as a proof-reader with a printer named Wilkins. There he took advantage of every opportunity to learn the printer's craft. It was about this time, *The Bath Journal* tells us, that he met his future wife. In order to support her after their marriage, he opened a printing house in the Strand. What might have become a prosperous business was cut short by the court action of other printers resentful that he had never served a trade apprenticeship. His case was heard at Westminster Hall in 1734, when he was fined. Shortly afterwards he was declared a bankrupt and one of his creditors asserted his rights to send him to a debtors' prison. Although the sum might be trifling,

23. John Nichols *Literary anecdotes of the eighteenth century*, vol. 2, 1812, p.93.

no prisoner could be freed until his total debt had been paid. Most of the inmates of the Marshalsea, the Fleet and the King's Bench prisons were debtors, and other London prisons had their quota. Jailers ruthlessly exacted payment for any amelioration of living conditions. Debtors often depended on the charity of family and friends to survive with a little dignity, but overcrowding and squalor had to be endured by most. One can be sure that Elizabeth Blackwell did all she could to improve the lot of her unfortunate husband. Her sole aim was now to obtain his release by paying his creditor.

Having drawn flowers since she was young, and learning that there was a need for an illustrated book on medicinal plants, she saw here a means of earning money. She submitted her project, supported by a selection of her paintings, to the influential physicians Sir Hans Sloane and Richard Mead and the apothecaries Joseph Miller and Isaac Rand. A 'public recommendation' of 1 October 1735 affirmed their approval and that of five other signatories, 'having seen a considerable number of the drawings from which the plates are to be engraved', and announced their 'good opinion of the capacity of the undertaker' (i.e. Mrs Blackwell). She gratefully accepted the invitation of Miller and Rand to make use of the resources of the Chelsea Physic Garden where they were employed.

In 1673 the Apothecaries' Company had leased just under four acres of land on the river at Chelsea to moor their official barge. In 1674 they enclosed it within a wall and a few years later they began transferring plants from their Westminster garden. Years of mismanagement were not resolved until 1722, when Sir Hans Sloane, Lord of the Manor of Chelsea, conveyed the land to them. Its finances now more secure, and managed by Isaac Rand and Philip Miller, the garden became known for its wide variety of plants, especially foreign introductions. When Linnaeus was in London in the summer of 1736 he visited the garden, hoping to acquire some of its choice specimens.

Also in 1736 Mrs Blackwell moved to no. 4 Swan Walk, abutting the eastern side of the physic garden, where she became a frequent visitor. Freshly cut plants were sent to her lodgings for her to draw. There she worked with a single-minded determination. Not only did she draw the 252 plants in the first volume of *A curious herbal*, which came out in July 1737, but as she was too poor to employ an engraver and colourist, she also etched her drawings and the accompanying handwritten text, and coloured the plates herself. Her imprisoned husband wrote brief descriptions of the plants depicted, their times of flowering and their medicinal uses taken from Joseph Miller's *Botanicum offinale* (1722), and used his knowledge of languages to give, wherever possible, equivalent plant names in Greek, Latin, French, German, Dutch, Italian and Spanish.

Mrs Blackwell dedicated her book to Richard Mead, 'as you was [*sic*] the first who advis'd its publication, and honour'd it with your name'. She presented a copy to the College of Physicians, which expressed its appreciation with a commendation and a present. The second volume, with 148 etched plates, was out either at the end of 1738 or early in 1739. She took great care to acknowledge everyone who had assisted, among them Sir Hans Sloane, for 'giving me the liberty to draw such foreign plants from your specimens (as were not to be had in England)', and Alexander Stuart, who had arranged for her earlier drawings to be exhibited at a plant-collecting excursion of the Company and for introducing her to Isaac Rand. Although the Chelsea Physic Garden was her principal source of plants, a few came from local nurserymen and she also adapted engravings in the *Hortus Indicus Malabaricus*.[24]

This 'monument of female devotion', as the *Dictionary of National Biography* described *A curious herbal*, appeared regularly in weekly parts, each with four plates, without fail for 125 weeks. Each part cost a shilling plain and two shillings coloured. A few copies were printed on large paper for one shilling and sixpence plain and three shillings with colouring superior to that in the cheaper version. The drawings are just adequate and the colouring a crude application of simple washes. Thomas Knowlton, a well-read and perceptive gardener, dismissed the book as 'a very silly stupid … thing' simply repeating what Gerard, Parkinson and other authors had written.

The University of Glasgow still has in its archives the records for the purchase of its copy of *A curious herbal*. Mrs Blackwell, or her agent, required a specified amount on receipt of the first delivery, a further payment after 300 plates had been delivered, and the balance when the work was finished.[25]

A curious herbal, sold by the bookseller Samuel Harding of St Martin's Lane, was an instant success. The author modestly expressed her astonishment at the 'favourable reception from the publick, both at home and abroad'. An advertisement in the *Country Journal, or Craftsman* for 6 May 1738 warned the public of the recent publication of 'a spurious and base copy', naming the engravers and printsellers involved: George Bickham, Philip Overton, John King, Thomas Bakewell, John Tinny, Samuel Simpson, Stephen Lye and Thomas Harper. Swift legal action by Alexander Blackwell thwarted this attempted piracy and each perpetrator was 'made to pay as dearly as he had done for the invasion of property'.[26]

John Nourse, a bookseller at the Lamb without Temple Bar, paid £150 for a third share of *A curious herbal* on 28 September 1737. This was an act of faith since only 80 parts had been published when he negotiated the deal. Nourse, a leading importer of foreign books, made the selling and publishing of scientific works his speciality. A shrewd businessman, he ensured that the book could not be produced without his consent by insisting on having a third of the copper plates.[27] An indenture between him and Blackwell, now living in Stanmore, on 19 February 1739 assigned the right to print, reprint, publish and sell *A curious herbal* for £319. 4s. 1d, which figure included a mortgage of £169. 4s. 1d owed by Alexander Blackwell to the bookseller. In October 1740 Nourse gave £75 for a sixth share of the copyright, which, combined with the third share he already held, gave him ownership of half the copyright. Nourse reprinted the book in 1739 and again in 1741

24. Vol. 2, plates 384, 391, 395, 400.
25. A.D. Boney 'A university purchase of Elizabeth Blackwell's Curious Herbal (1737–1739)'. *The Linnean*, vol. 3, part 3, 1987, pp.35–38.
26. *Gentleman's Magazine*, September 1747, p.425.
27. Add MS 38729, ff.34–48. British Library.

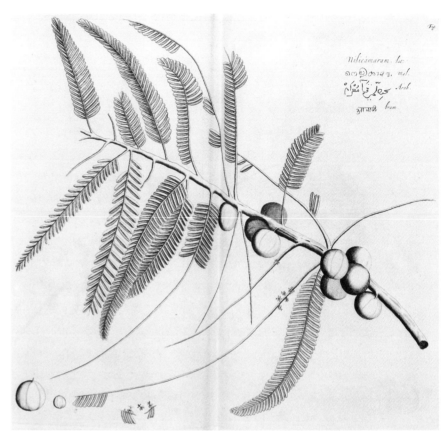

Elizabeth Blackwell *A curious herbal*, London 1737–39.
LEFT Vol. 2, plate 400. Emblick Myrobalan (*Phyllanthus emblica*), based on a drawing (RIGHT)
from Van Reede's *Hortus Indicus Malabaricus*, vol. 1, fig. 38. B.L.452.f.2 and B.L.39.453.f.7.

OPPOSITE
Elizabeth Blackwell *A curious herbal*, London 1737–39. B.L.452.f.2.

TOP LEFT Vol. 2, plate 358. Dragon-tree (*Dracaena draco*). Probably one of the tender plants grown
under glass in the Chelsea Physic Garden. She took her drawing of a mature tree from a work by Clusius.
'… esteemed restringent, drying and binding … It also fastens loose Teeth and stops the Bleeding
of the Gums & helps the Scurvy in them.' Sold in shops as 'Dragon's blood'.

TOP RIGHT Vol. 2, plate 360. Scythian lamb, described by Blackwell as 'a moss that grows upon the
roots of a fern'. It is the rhizome of a fern (*Cibotium barometz*), the Tartarian lamb of early travellers.
Blackwell probably copied the specimen described by Sir Hans Sloane: 'It seems to be shaped by art
to imitate a lamb, the roots or climbing part being made to resemble the body, and the extant footstalks
the legs' (*Philosophical Transactions of Royal Society*. Abridged edition, vol. 20, 1698, pp.345–46).

BOTTOM LEFT Vol. 2, plate 409. Columbine (*Aquilegia vulgaris*). 'Good for sore mouths and
inflammations of the jaws and throat.' Compare this engraving with the same plate in the German edition.

BOTTOM RIGHT Elizabeth Blackwell *Herbarium Blackwellianum emendatum et auctum*,
Nuremberg 1747–73. Vol. 3, plate 409. *Aquilegia vulgaris*.
The main difference between the English and German editions of this book
is the addition of floral dissections to the latter. B.L.735.K.9

The Dragon-Tree. Draco Arbor.
Eliz. Blackwell delin. sculp. et Pinx.
1. The Tree in ye Physick Garden in Chelsea.
2. A Copy of the Tree from Clusius.
3. Flowers.
4. Berries.
5. one Berry of its full Size.

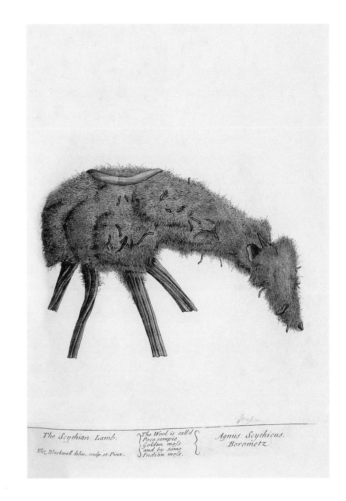

The Scythian Lamb. The Wool is call'd Poco sempie Golden moss and by some Indian moss. Agnus Scythicus. Borometz.
Eliz. Blackwell delin. sculp. et Pinx.

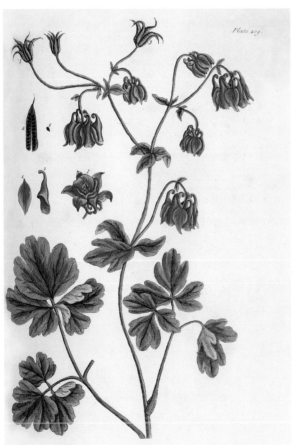

after advertising it in the London and country papers. On 5 December 1745 Alexander Blackwell, now resident in Ollerstad, near Gothenburg, in Sweden and describing himself as 'Doctor in Physic', gave power of attorney to his wife in any future negotiations with Nourse. In April 1747 Elizabeth Blackwell surrendered their remaining half share in her book, all unsold copies and the copper plates to Nourse as a final settlement of outstanding debts. The book was a good investment. 'The work still continues in such esteem as to keep up its original price of six or seven guineas and 10 on large paper in the modern sale catalogues,' observed the *Gentleman's Magazine* in 1806 (part 2, p.1101).

Its popularity was not confined to Britain. Christoph Trew, the Nuremberg promoter of natural history books, purchased a copy – probably the large paper issue. He liked the work well enough to undertake a German edition, a decision that surprised Peter Collinson, who wrote to him about it on 18 January 1753:

> But my dear friend wee are all surprised, considering your great practice as a physician, and the many other works that are under your hands, & require so much of your time, that you could find leisure, or think it worth your notice, (unless you intend it as a complement to the fair sex) to bestow so much time & pains to republish Mrs Blackwells Herbal, a work without order or methode, & of no esteem here and only encouraged as an act of charity to the poor woman.[28]

Trew ignored this advice, resolved to expand the meagre English text and commissioned an artist-engraver, Nicholaus Friederich Eisenberger, to copy and publish the plates. In 1747 Eisenberger announced details of the work's forthcoming publication in quarterly parts, each with 16 (later reduced to 15) plates.

The problems in producing a serial publication were no different in Germany. The first part duly appeared in the summer of 1747, but the next, scheduled for the autumn, was delayed until the following spring. The blame lay with Trew, who indulged in excessively long notes. By the end of 1749 only 90 plates had been issued and subscribers were getting restless. Trew's interest had been diverted to writing the text of *Plantae selectae*, a selection of Ehret's botanical drawings which he owned. An impatient Eisenberger sent out the plates without Trew's text and exasperated subscribers defected. In 1752 Trew handed over the responsibility for the text to a couple of young botanists, but could not resist revising their contributions.

Even with editorial help the text still trailed behind the plates, incurring financial losses for Eisenberger. In a notice accompanying the 38th part in December 1759 he appealed to defaulting subscribers to pay their arrears and to pledge themselves to support an additional 100 plates. The Seven Years' War (1756–63) provided a convenient excuse for its protracted publication, but finally, in January 1765, the last of the 500 plates copied from *A curious herbal* were despatched. Now a sick man, Eisenberger was unable to finish the extra plates before his death in 1771. That task fell to Professor G.G. Ludwig, who

28. Peter Collinson letters. Linnean Society.

had long been associated with the work; launched in 1747, it took a quarter of a century to complete. The 615 plates in *Herbarium Blackwellianum emendatum et auctum* are on the whole superior to those in *A curious herbal*, better engraved and coloured.

And what happened to restless and impetuous Alexander Blackwell after two years in a debtors' prison? Conflating the information in his obituaries in *The Bath Journal* and the *Gentleman's Magazine* and the *Dictionary of National Biography*'s entry, it would appear that he spent some of his leisure time studying medicine – probably little more than reading textbooks on the subject. He also turned his attention to methods of land reclamation and utilization. We learn that this new enterprise led to the Duke of Chandos putting him in charge of improvements on his estate at Canons in Middlesex. He held the post briefly, leaving 'under circumstances not explained but apparently little to his credit'.[29] While still in the Duke's service, he published anonymously *A new method of improving cold, wet, and barren lands: particularly clayey-grounds* (1741).

Our 'indomitable adventurer', as the *Dictionary of National Biography* describes him, surfaces in Sweden in 1742. It seems that the Swedish ambassador in London, impressed by Blackwell's book, sent a copy to the Royal Court in Stockholm. Having accepted an invitation to visit Sweden as an agricultural adviser, Blackwell left his wife and child behind, trusting that they might join him later. He was warmly welcomed and appointed director of a model farm at Allestad. When the King fell ill, Blackwell found an effective remedy and was rewarded with the post of physician in ordinary to His Majesty.

As one tries to make sense of Blackwell's Swedish episode one is faced with conflicting accounts. Another version comes from Linnaeus himself, who met Blackwell at Allestad in 1747. Linnaeus gives an account of this meeting and what he knew of Blackwell in his *Nemesis divina*, a personal document of paternal advice left for his son. He uses the fate of Blackwell and other 'miscreants' as evidence of retributive justice. But it must be remembered that the Swedish naturalist disliked Blackwell – 'a bold, ignorant atheist,' he called him – and that his report would be prejudiced.

Linnaeus's informant and friend Jonas Alströmer was respected as the 'father of Swedish arts and crafts'. When Alströmer requested an economist from England, Blackwell was selected, according to Linnaeus. Despite being well received, he turned out to be a treacherous guest. Accidentally opening one of Blackwell's letters, Alströmer discovered that he was recommending his assassination and that of other politicians to make way for an English heir-apparent to the Swedish throne. The Council of State resolved to be rid of Blackwell when this incriminating letter reached it. Linnaeus portrayed him as a thoroughly evil person. He accused Blackwell of having an affair with the wife of a commission agent in Stockholm. When her husband became ill, Blackwell treated him, with fatal consequences. The President of the College of Commerce was another victim of Blackwell's medicine. Incompetence, rather than murder, was the more likely cause of death.

29. *Dictionary of National Biography.*

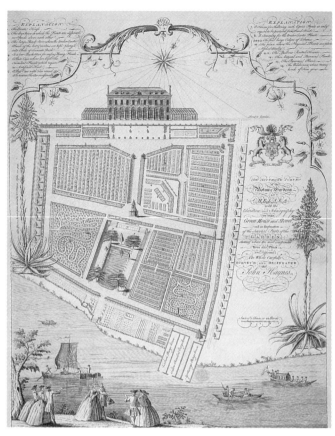

Chelsea Physic Garden. Surveyed and plan drawn by John Haynes, 30 March 1751.
This garden provided Elizabeth Blackwell with many of her specimens to copy. B.L.K.Top 28.4.p.

The *Dictionary of National Biography* presents another scenario. When the farm at Allestad deteriorated under his supervision, Blackwell's precarious position tempted him into political intrigue. Ruled by a weak monarch, the country was unstable; the two opposing parties, symbolized by headgear, were the 'Hats' supported by France and the 'Caps', the favourites of the British. In March 1747 Blackwell gave the King a secret communication from the Queen of Denmark, George III's daughter, offering a substantial bribe if the succession to the throne were altered. The King told his ministers, Blackwell was arrested and without a public trial sentenced 'to be brok alive on the wheel, his heart and bowels to be pulled out and burnt, his body quartered, and his head fix'd upon a pole'.[30] The sentence was commuted to a tidy and swift beheading, which took place in Stockholm on 29 July 1747. A modern historian has called it 'a judicial murder'. Blackwell's life had been a succession of disappointments and disasters, and he was almost certainly innocent of the many charges laid against him, a victim of two feuding political parties.

We do not know what happened to his long-suffering wife, Elizabeth, who died in October 1758 and was buried in Old Chelsea Church, having outlived her husband by eleven years.

30. *Gentleman's Magazine*, June 1747, p.298.

8 Banks's florilegium

The crews of ships on voyages of discovery usually included someone capable of sketching views of new lands and their inhabitants. The former buccaneer William Dampier, an experienced sailor and navigator, commanded H.M.S. *Roebuck* in 1699 on a survey of the north-western coast of Australia, New Guinea and the South Pacific. In his own words: 'Where I found a harbour or river I would land and seek about for men and other animals, vegetables, minerals, etc., and having made what discovery I could, I would return home by way of Tierra del Fuego.' The Admiralty wanted not only to know what lay to the south-east of the Dutch East Indies, but also to gain some idea of the natural resources of the region. Dampier had a talented sailor on board to draw vegetation and wildlife, and a selection of his drawings were reproduced in *A voyage to New-Holland in the year 1699* (1703) together with coastal profiles of islands, presumably by the same artist. Being an amateur naturalist, Dampier added informative notes on the fauna and flora. His voyage was possibly the first to have scientific exploration as one of its objectives.

In 1766 Louis-Antoine de Bougainville was appointed by the French Government to search for territories in the Pacific fit for colonization. One of the two ships commissioned for the voyage, the *Etoile*, had on board an astronomer, to pursue investigations into methods of calculating longitude, and a doctor/naturalist, Philibert Commerson. The latter made many competent drawings and accumulated a substantial collection of specimens, but Bougainville's account of the voyage concentrated on the sensational and ignored this scientific data. A few years later a Royal Naval vessel, H.M.S. *Endeavour*, demonstrated how a scientific voyage should be organized. Neither expedition published a scientific report of its discoveries.

The British and French were rivals in the race to explore the Pacific and to claim *Terra Australis Incognita* once it had been found. John Byron's expedition in 1764 was unsuccessful, although he did survey the Falkland Islands, which both nations coveted as a strategic base. In 1767 Samuel Wallis discovered Tahiti, which he named King George's Island. In 1768 the British Government decided to send a ship there to observe the transit of the planet Venus over the Sun on 3 June 1769. At the same time British astronomers would carry out similar observations in Spitzbergen and Hudson's Bay. This was the official mission, but the commander of the expedition would receive secret instructions to search also for the legendary southern continent.

Alexander Dalrymple, Fellow of the Royal Society, an authority on Pacific exploration and a committed believer in the existence of the elusive continent, considered himself to be the best qualified to lead this scientific survey. The Admiralty, however, refused to accept a civilian in charge of one of its ships. It was not until April 1768 that a middle-aged seaman, James Cook, was nominated as candidate for the post. After five years on a collier trading between Newcastle and London he had joined the Royal Navy, charted the St Lawrence river in Canada, and surveyed the Newfoundland and Labrador coasts. The Commander-in-Chief of the North American Station recommended him as 'well fitted'. Cook accepted the assignment and selected an east coast collier of 366 tons as suitable for conversion. This sturdy little vessel, reinforced by an extra layer of planks and given new masts, was renamed *Endeavour*.

Had it been known that Joseph Banks and his entourage were to join the *Endeavour*, a larger ship might have been chosen. Banks was then a young man of 25, newly elected a Fellow of the Royal Society, with the financial resources to indulge his passion for botany. In 1766 he had spent nine months on a naval ship on duty off Newfoundland and Labrador, an experience which had given him a taste for foreign travel and the urge to go where no naturalist had been before. The *Endeavour* afforded that opportunity. Ever persuasive and persistent, he got the Royal Society to petition the Admiralty on his behalf. It was always understood that he would meet the costs of his participation. When friends reminded him of the risks and advised instead a conventional tour of continental Europe, he is reputed to have retorted: 'Every blockhead does that; my grand tour shall be one around the whole globe.' His companions were Daniel Solander, a former pupil of Linnaeus, as botanist, Hermann Spöring as scientific secretary, Sydney Parkinson as natural history draughtsman, Alexander Buchan as landscape painter, and four servants.

With every inch of space precious, Cook must have been dismayed by the amount of equipment Banks brought on board: microscopes and other scientific instruments, bags, baskets and bottles to store specimens, nets and trawls for catching marine life, insect traps, guns and pistols for shooting animals and birds, vasculums for collecting plants, beeswax and a stock of salt for preserving seeds, paper and paints for the artists, and a library of essential natural history books. 'No people ever went to sea better fitted out for the purpose of natural history, nor more elegantly,' wrote John Ellis to Linnaeus. *The Endeavour* sailed from Plymouth in August 1768 with 94 men on board.

Central to Banks's plan for a floral record of the voyage was an artist capable of working under pressure in cramped quarters on board ship, adjusting to the vagaries of climate, all the time maintaining a reasonable output of drawings. He was most fortunate to meet Sydney Parkinson (*c*.1745–71), who, in the event, amply satisfied these requirements. The son of a Quaker brewer in Edinburgh, Parkinson had been apprenticed to a woollen draper. According to his elder brother Stansfield, he had always taken 'a particular delight in drawing flowers, fruit and objects of natural history'. No evidence exists that he ever received any formal art training, but his skill and sense of design suggest some tuition. Avril Lysaght made out a case for his having been a pupil of William de la Cour, the French proprietor of a school of drawing and design in Edinburgh.

It is not known when Parkinson moved to London with his mother, but in 1765 he exhibited a flower painting on silk at the Free Society of Artists, and the following year two flower drawings. It was a stroke of luck when he met a fellow Scot and Quaker, James Lee, a nurseryman at Hammersmith. They became friends and Parkinson instructed Lee's daughter Ann, then thirteen years old, who clearly had a talent for painting flowers. It was probably Lee who introduced him in 1767 to one of his customers, Joseph Banks, who commissioned him to draw a selection of his Newfoundland specimens and to record birds and mammals in the Governor of Ceylon's collection (these were used by Thomas Pennant in his *Indian Zoology*). Satisfied with this evidence of his work, Banks engaged Parkinson at an annual salary of £80 and full board as his natural history draughtsman during the voyage. Parkinson was 23 years old, now on the threshold of a promising career.

Cook magnanimously allowed Banks and Solander to work in his great cabin and there Parkinson drew whatever objects they brought him, getting accustomed to the motions of the ship as it headed towards Madeira. But the fifth day out it pitched so much that the artist gave up. Banks and Solander gathered some 400 plants during the five days the *Endeavour* reprovisioned at Madeira. From those selected for him to draw, Parkinson completed sixteen watercolours. As the ship made for Rio de Janeiro he tested his versatility by sketching the corpse of a young shark.

The Portuguese authorities at Rio refused permission for the crew to land, apart from a party loading supplies, but Banks was not defeated by this prohibition. 'Our few botanical collections have been made by clandestinely hiring people; and we have got them on board under the names of greens for our table. Now and then we have botanised in the bundles of grass that have been brought to or goats and sheep.' Banks, Solander and Parkinson daringly sneaked ashore on one occasion, returning with armfuls of flowers of which 30 of the best were painted by Parkinson.

Tierra del Fuego at the tip of South America lived up to its inhospitable reputation. The thickly wooded hills were almost impenetrable, the natives were, wrote Cook, 'as miserable a set of people as are this day on earth', and Banks lost two of his servants in a blizzard. Half of the 150 species of plants found there had been recorded in watercolours by the time the *Endeavour* sighted the Society Islands in the Pacific.

On 13 April 1769 they anchored off Tahiti to a cautious welcome from the islanders. The euphoria Banks must have felt was tempered a few days later by the death of his landscape artist, Alexander Buchan, brought on by an epileptic fit. He confided to his

Pencil drawing by Sydney Parkinson of H.M.S. *Endeavour* beached on the Endeavour River, Australia, after striking a coral reef in June 1770. B.L.Add MS 9345, f.57.

journal that 'his loss to me irretrevable [*sic*], my airy dreams of entertaining my freinds [*sic*] in England with the scenes that I am to see here are vanished'. Buchan's unexpected demise meant more work for Parkinson, now required to record the lifestyle of the Polynesians as well as their lush vegetation.

Flies distracted him as he worked. 'We can scarce get any business done for them,' wrote Banks in his journal (22 April 1769), 'they eat the painters colours off the paper as fast as they can be laid on, and if a fish is to be drawn, there is more trouble in keeping them off it than in the drawing itself.' To deter them a huge net was draped over Parkinson and his table and chair, with an insect trap to attract any intruders that got through.

Besides sketching events and ceremonies for Banks, Parkinson added another 113 finished flower paintings to his portfolios. With so many plants being brought to him he was compelled to be selective, to sketch more in outline and to finish fewer. Merely a flower and a leaf would be coloured on some pencil drawings; others would be annotated with terse colour notes on the reverse side of the drawing: 'the small petals dirty white', 'calyx green yellow ting'd & freckled wt red', for example. His colour code was imprecise: 'bright green', 'dark red', 'pale yellow', 'stalks sordid brown' were phrases open to various interpretations, but this shorthand was for Parkinson's sole convenience.

The astronomical observations accomplished, the *Endeavour* left in August bound for New Zealand. Back at sea again Banks and his two companions reverted to the comfortable routine they had established during the voyage. Mornings were reserved for consulting books in their well-stocked library. From late afternoon Banks and Solander sat at the long table in Captain Cook's cabin, describing the plants collected and selecting those for Parkinson to draw. He faced them, where they could make comments on his composition and accuracy. All three worked swiftly before the plants wilted and died. Spöring filed their botanical notes in geographical sequences, forming the basis of rudimentary regional floras. He had also taken some of the pressure off Parkinson by sketching landscapes, people and animals.

Banks and Solander faced hostile Maoris in New Zealand to gather a bumper harvest of plants, the biggest so far on the voyage. Parkinson was overwhelmed: out of more than 200 rapid sketches he could only manage to produce 30 watercolours.

Eastern Australia proved his greatest challenge. The vegetation was so prolific on one stretch of the coast that Cook named it Botany Bay. Parkinson rose to the occasion. Banks's journal for 12 May 1770 approvingly noted that 'in 14 days, just one draughtsman has made 94 sketch drawings, so quick a hand has he acquired by use'. Parkinson worked well into the night by the light of an oil lamp, eventually producing 408 outline drawings and three finished ones.

After a most successful voyage Cook now headed for home, calling at the Dutch settlement at Batavia with its well-made roads, canals, elegant villas, churches and commercial buildings. It was, however, a most unhealthy place, whose European residents feared an early death from malaria or dysentery. The crew of the *Endeavour* soon succumbed. The ship's doctor died, Solander became critically ill, Banks was very unwell. Spöring and Parkinson died within a couple of days of each other in January 1771. Cook had protected most of his crew from scurvy but he had no remedy for tropical diseases. After essential repairs to the ship, he recruited seamen to replace those who had died, and left Batavia as soon as possible.

According to Harold Carter's definitive life of Sir Joseph Banks, Sydney Parkinson produced 942 botanical drawings, 269 of them watercolours, the remainder pen or pencil sketches. In addition he left zoological subjects, landscapes and ethnographic studies. Although not an outstanding botanical artist, he worked diligently and accurately, as Banks confirmed in a letter to John Fothergill:

> *Now as S. Parkinson certainly behaved to me, during the whole of his long voyage, uncommonly well, and with unbounded industry made for me a much larger number of drawings than I ever expected, I always did and still intend to shew to his relations the same gratitude for his good services as I should have done to himself.*[31]

The *Endeavour* docked at Dover on 17 July 1771. Only 41 of the ship's original company of 94 had survived; five of Banks's party had died. The ecstatic welcome given to Banks, now a celebrity, eclipsed any recognition of Captain Cook's superb seamanship. His zoological specimens exceeded 1000 species; his plant specimens, estimated at about 30,000, included many new to botanists. Society flocked to admire the work of his artists and the curiosities he had collected. He and Solander were graciously received by George III. Linnaeus hailed him as 'immortal Banks'.

It has to be admitted that this public adulation went to his head. When two ships were being fitted for another voyage to the Pacific, again under Cook's command, Banks naturally thought he would be part of it. In anticipation he selected fifteen people: Solander, of course, four artists and ten servants, including two horn players. The Navy Board, fearing that the structural alterations Banks demanded might affect the ship's stability, strongly objected. Banks withdrew in pique and went to Iceland instead in July 1772. When he returned to London at the end of the year he gave serious consideration to a scientific account of his *Endeavour* voyage.

The general public as well as scientists and scholars eagerly awaited its publication. Linnaeus, motivated by impatience, was concerned that, if he delayed, his collections

31. S. Parkinson *Journal of a voyage to the South Seas*, 1773, pp.4–5.
32. 22 October 1771, J. E. Smith *A selection of the correspondence of Linnaeus*, vol. 1, 1821, pp.267–69.

might become 'the prey of insects and of destruction'. He conveyed his anxiety to J.E. Smith in October 1771:

> *Do but consider, my friend, if these treasures are kept back, what may happen to them. They may be devoured by vermin of all kinds. The house where they are lodged may be burnt. Those destined to describe may die ... I therefore once more beg, nay I earnestly beseech you, to urge the publication of these new discoveries.*[32]

With his Icelandic excursion behind him, Banks launched his project in January 1773. He re-engaged the three artists whom he had chosen to join him on Cook's second voyage but who instead had accompanied him to Iceland: the two Miller brothers – James and John Frederick – and John Clevely. They were set to work to produce watercolours based on Parkinson's unfinished drawings, guided by his colour notes and specimens in Banks's herbarium. Before they left his employ in 1776, J. F. Miller had completed 99 watercolours, his brother 85 and J. Clevely 26. An obscure artist, Frederick Polydore Nodder, who replaced them, eventually added another 271.

Assured by a gradual accumulation of drawings, Banks considered the next stage – the selection of engravers and available methods of engraving: line, etching, mezzotint

Serjania cuspidata, collected by Banks in Brazil. Painted by Sydney Parkinson in 1768.
Natural History Museum, London.

and stipple. Some of the best British botanical engravers, such as William Kilburn, John Sebastian Miller, Francis Sansom and James Sowerby, were already committed to other books. Banks looked to Europe for recruits. He examined trial proofs from a Berlin engraver, but his fees were unacceptable. Gerhard Sibelius was found through a Dutch friend. Between 1773 and 1784, 18 engravers were employed, the most productive being Daniel Mackenzie with 251 plates, G. Sibelius with 195 and Gabriel Smith with 117. The quality of Mackenzie's work was not diminished in any way by his higher output. He was undoubtedly the best of the team. His sensitivity can be seen in his rendering of Francis Bauer's delicate erica drawings in *Delineations of exotick plants … at Kew* (1796–1803) and the superb compositions in A. B. Lambert's *Genus Pinus* (1803–24). He was the principal contributor to Roxburgh's *Plants of the coast of Coromandel* (1795–1820).

It is believed that Banks looked at the possibility of printing from coloured copper plates rather than hand-colouring the prints themselves. He may have considered the *à la poupée* method of dabbing coloured inks on the plates with a small cloth bundle, but he opted for the engraver's technique of conveying tonal values by parallel lines and cross-hatching, thereby producing a subtle graduation from rich black to pale grey.[33]

Solander's death in 1782 may temporarily have delayed progress, but his successor, Jonas Dryander, supervised the artist and the engravers. In a letter to J. Alström in November 1784 Banks forecast completion of the work within a few months. Not all of Parkinson's outline drawings had been expanded into finished watercolours when Nodder left that year.

Banks now had 753 engraved copper plates and the completed text for the early part of the voyage at his disposal. One can only speculate why he did not proceed to publication. Calculations based on the fees commanded by contemporary artists and engravers indicate that he must already have spent £4500. Another £7500 might have been needed to print and bind the work, envisaged to appear in 14 or 15 parts. The general economic depression and, in particular, a reduction in his income from his Lincolnshire estates may have prompted a prudent withdrawal. The disappointing sales of a work Banks much admired – William Curtis's ambitious *Flora Londinensis*, started in 1775 – may also have made him hesitate. But financial caution was never the excuse he gave to friends and enquirers. 'Je n'ai pu faire que tres peu de progres dans mon propre ouvrage,' he admitted to M. van Marum in May 1791, blaming his procrastination on his involvement in the management of the royal gardens at Kew.

There may be another reason why his project collapsed. Banks published nothing of consequence himself; he was, in fact, a reluctant author. His book might have materialized had Solander been alive to finish the text and supervise its production. Banks always regarded himself as a catalyst, a person with ideas and initiative, one who encouraged others to carry them through. Without his backing, the *Delineations of exotick plants … at Kew* and *Plants of the coast of Coromandel* would never have been published. Now all

33. Its effectiveness can be judged in the monochrome plates printed by the Lion and Unicorn Press in 1973.

he had to show for his efforts and expense were three sets of trial proofs of the engraved plates; one set he retained and the rest he distributed among other institutions and to botanical friends.

On Sir Joseph Banks's death in 1820 his collections, including the *Endeavour* drawings and the engraved copper plates, were bequeathed to Robert Brown, his curator and librarian, on the understanding that on his death they would pass to the Trustees of the British Museum. Brown transferred them in 1827 to the British Museum, where he was appointed a keeper.

In June 1890 William Carruthers, Keeper of Botany at the British Museum (Natural History) – the natural history collections had been transferred from the British Museum to South Kensington in 1881 – proposed the publication in four volumes of the plates depicting the plants collected on Cook's first and second voyages in an edition of 300 copies. He estimated that the cost would be £900, of which the booksellers Dulau and Company agreed to give £400 for half the imprint. Another £100 might be raised from the sale of the copper plates, presumably for scrap.

The survival of any copper plates has always been unpredictable. Far too often they were melted down for their value as copper. About 1880 Quaritch the bookseller reported that Sibthorp's *Flora Graeca* 'coppers were recently sold as metal'. Robert Thornton deliberately destroyed those for his *Temple of Flora* to meet the requirements of his lottery. Kew Gardens disdained the offer of the plates of *Plants of the coast of Coromandel*, assessing them to be worthless.

The Trustees of the British Museum deferred making a decision on Carruthers's proposal, which was resurrected by G. Murray, Carruthers's successor, in November 1898. Eventually lithographs were made from 318 of the original set of trial proofs and published as *Illustrations of Australian plants of Captain Cook's voyages round the world in HMS Endeavour in 1768–77* (1900–05).

The copper plates which had miraculously survived for so many years narrowly escaped destruction on 10 September 1940 during an air raid that severely damaged the Department of Botany in the Natural History Museum in South Kensington. Salvaged, they were stored in piles, unprotected and neglected, until 1961, when the Royal College of Art, having cleaned a few sample plates, discovered that they could still yield excellent impressions. It was decided in 1968 to select 30 of them on the basis of the 'botanical and historical interest of the plants portrayed' for printing on a hand press at the College's *Lion and Unicorn Press*. One hundred sets of *Captain Cook's florilegium* (a misleading title, but the name of Cook means more to the public than that of Banks) were published in 1973.

While he was on business at the Royal College of Art, Mr J. Studholme of Alecto Historical Editions heard about this large cache of copper plates. Enquiries were made at the Natural History Museum, some trial proofs were pulled, and in 1980 a prospectus announced a joint venture between the Museum and the publisher to issue 100 sets of coloured prints taken from 738 plates (15 had been lost in the past two centuries).

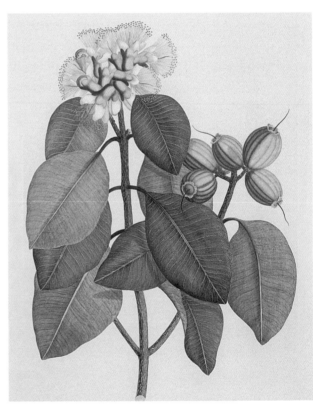

Syzgium suborbiculare, now *Eugenia suborbiculare*.

TOP LEFT Pencil drawing by Sydney Parkinson, made on Lizard Island, Queensland, 1770.
Colour note on reverse: 'flower & artery white'. Natural History Museum, London.

TOP RIGHT Painted by F.P. Nodder in 1777, based on Parkinson's drawing.
Natural History Museum, London.

BOTTOM LEFT *Captain Cook's florilegium*, Lion and Unicorn Press, 1973. Plate 19.
Printed from the original copper plate. A good example of the engraver's skill in rendering
the delicacy of the flower, the texture of leaves and subtle highlights.

BOTTOM RIGHT Printed by Alecto Historical Editions from original
copper plates and colour-printed *à la poupée*.

A finished print being pulled off the press at Alecto Historical Editions.
Up to ten colours may be worked directly into the single plate before each print is pulled.
Additional details are added in watercolour.

A special studio was equipped in a disused warehouse and extra staff were taken on for this daunting project. All the coppers were meticulously cleaned to remove acid stains that had migrated from paper wrappers and also hardened printers' ink from the original printing of proofs. which had clogged finely engraved lines. Burnishing and a coating of chrome followed to protect them against wear during printing.

Knowing that Banks had investigated a technique for colouring the plates, it was thought appropriate to issue them in colour. Colouring by hand of thousands of black and white impressions was considered too costly and protracted a process. Instead, the application of colour on the plates *à la poupée* was chosen. At the outset no more than six or seven colours were added to any of the plates, but with perfected skills up to eighteen were later applied. Sir Joseph Banks would surely have applauded the impressive achievements of a dedicated team of craftsmen and artists who, over a decade, never deviated from impeccable standards in print-making. The precious copper plates, now protected in acid-free wrappings, are safely back in the Natural History Museum.

9 William Curtis

When Sir Joseph Banks was looking for a bibliographical model for his illustrated account of the *Endeavour* voyage, he chose the *Flora Londinensis*, a work which, he declared, 'unites elegance with perspicuity'. Its author, William Curtis, was, like Banks, an entrepreneur, but without the good fortune that usually attended Banks's ventures.

Curtis was born on 11 January 1746 at Alton in Hampshire, the son of a tanner and a Quaker. A precocious interest in the local fauna and flora probably influenced his parents' decision to apprentice him to his grandfather, the town's surgeon/apothecary. He subsequently worked for two London apothecaries to broaden his experience. For a while he had his own practice, but in the words of Sir James Edward Smith, 'the streetwalking duties of a city practitioner but ill accorded with the wild excursions of a naturalist: the apothecary was soon swallowed up in the botanist, and the shop exchanged for a garden.'[34] In 1771 Curtis purchased an acre of land in Lambeth, eventually converting it into a garden of British plants. The same year witnessed the publication of his first book, an unpretentious pamphlet describing several methods of collecting and preserving insects. He followed this with an English version of Linnaeus's *Fundamenta entomologiae* (1772). He had now experienced the pleasures of publishing, but in the meanwhile his growing reputation as a naturalist led to his appointment in 1773 as Demonstrator of Botany at the Chelsea Physic Garden. His duties were not particularly demanding for a young man unwilling to remain an obscure apothecary. Four years later he resigned from this post to devote more time to publishing and other activities. *Linnaeus's system of botany*, based on lectures Curtis had given to students, appeared in 1777. In 1778 he issued his *Proposals for opening the London Botanic Garden*, where subscribers would be introduced to plants useful in agriculture, medicine and cooking.

His main preoccupation, however, was the publication of the *Flora Londinensis*, which he hoped would make his fortune and enhance his botanical reputation. An anonymous *Catalogue of the plants growing wild in the environs of London* (1774) was attributed by Jonas Dryander, Banks's librarian, to William Curtis. Although this attribution is in doubt, the catalogue may have suggested to Curtis the concept of the *Flora Londinensis*, which initially aimed to describe and illustrate 'such plants as grow wild in the environs of London' (title-page) – that is, within a ten-mile radius of the capital. Subscribers would receive it in a series of parts, each one containing six plates and accompanying text. Each part would cost two shillings and sixpence uncoloured, five shillings coloured, or seven shillings and sixpence for the few with superior colouring. All the plants were to be depicted life-size, or, if too large, by a stem or branch.

34. A. Rees *Cyclopaedia*, vol. 10, 1819.

Curtis engaged William Kilburn (1745–1818), whose flower paintings he had seen displayed in print shops, as artist. Kilburn, who had been apprenticed to an Irish textile designer before going to London, was Curtis's sole artist from 1775 until about 1777/8, when he became manager of a Surrey calico factory. He was succeeded by James Sowerby and Sydenham Edwards. Thomas Milton, who engraved the plates for about a year, was replaced by Francis Sansom. During the 23 years it took to issue the 72 parts of the *Flora Londinensis* a team of colourists was supervised by William Graves, an expert in colouring natural history works. Subscribers complained about the variable quality of the hand-colouring. Dr William Hird complained that 'the colouring of every plant of the first number in my possession is by no means equal to the specimen [copy] Cousin Nancy Freeman left with me'.

Other subscribers were annoyed by the erratic appearance of the parts. 'I lament only that the numbers do not come out oftener,' wrote the botanist John Lightfoot. 'Is it not possible for you to hasten your *Flora Londinensis?*' enquired an exasperated Thomas Cullum, who offered Curtis some unsolicited advice: 'If you will engage to publish your *Flora* in any limited time I am ready to subscribe [to] any proposition in advance, this is the only way you can ever think of getting thro' such an undertaking.'

Poor planning perhaps, but inadequate financial resources certainly, were responsible for some of these delays. Curtis's patrimony, earnings from lecturing, subscriptions from his garden and, from 1787, profits from his *Botanical Magazine* all went into the project. He begged loans from friends. In October 1778 John Coakley Lettsom agreed to lend him £500 in instalments to 'assist thee in the prosecution of thy work, which I consider as a National honour' with no intention of reclaiming the debt. Lord Bute was another generous supporter. Curtis had evidently allowed too small a margin of profit when he originally costed this publication. He blamed the loss of William Kilburn for some lapses in his publication schedule, but in truth he dissipated his time in too many pursuits: managing a botanic garden, leading botanical forays around London, editing a new magazine, launching in 1786 *Assistant plates to the materia medica* (which collapsed after only two issues) and issuing occasional leaflets. He clearly lacked discipline and focus.

In 1792 he speculated on *An abridgement of the Flora Londinensis* in a smaller format, but after five issues, amounting to 36 plates, it was abandoned. In 1798 the *Flora Londinensis* itself ceased publication. It had accumulated 432 plates which conflated into six fascicles making up two folio volumes. The first volume was dedicated to Lord Bute and the second to his other principal benefactor, J.C. Lettsom. As Curtis was in the habit of revising his lists of subscribers it is not known how many he had, but 320 is near the mark. Among his customers were a number of distinguished peers, about 30 surgeons and physicians, 21 apothecaries, 11 nurserymen and 27 clerics. The Duchess of Portland, J.C. Lettsom and William Baker of Berkeley Square each subscribed to two sets of the work.

The *Flora Londinensis* with competently drawn and engraved figures never deserved its failure. One factor could have been the difficulty in consulting it. Neither the pages

LEFT William Curtis *Flora Londinensis*, 1775–98. Vol. 1. *Veronica chamaedrys or Germander speedwell.*
Drawn and engraved by *William Kilburn.* B.L.1823.e.4.

RIGHT William Curtis *An abridgement of the Flora Londinensis*, London 1792. Smaller versions of larger
works are seldom successful. The quarto edition of Thornton's *Temple of Flora* was another failure.
B.L.1509/659.

nor the plates are paginated, making precise citation impossible. Curtis had intended
that they should be arranged according to Linnean classification on the completion of the
work. Its protracted publication certainly deterred subscribers.

When William Curtis died in 1799 he bequeathed his estate, including the *Flora
Londinensis*, to his widow and to his daughter, who in 1801 married Samuel Curtis, her
father's cousin. In 1802 her husband agreed to pay the executors of William Curtis
£600 for the stock of the *Flora Londinensis* and its copper plates 'for the purpose of selling
or continuing the publication thereof as he might think proper'. Failing to receive any
payment from Samuel Curtis, the executors repossessed the *Flora Londinensis*. In January
1815 George Graves, son of the colourist William Graves and the husband of William
Curtis's niece, became part-owner of the *Flora Londinensis*. Under his supervision a new
edition was published in five volumes between 1817 and 1828. The number of engravings expanded to 658 and a solitary lithograph. The majority of the additional plates and
text were the work of William Jackson Hooker, Professor of Botany at Glasgow University.

The new edition was in no way an inferior production, but it, too, failed to attract
enough subscribers. George Graves had rashly bought out his three partners for £227,

 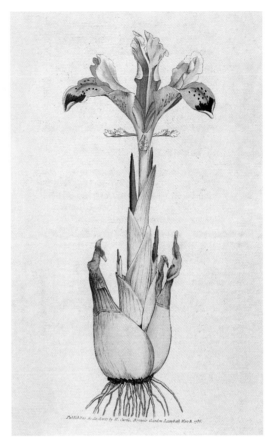

LEFT Watercolour of *Iris persica* by James Sowerby.
Painted for the first issue of the *Botanical Magazine* in February 1787. Fitzwilliam Museum, Cambridge.

RIGHT *Botanical Magazine*, London 1787. Vol. 1, plate 1. *Iris persica*. Although this engraving is competently hand-coloured, the colours do not correspond exactly with those in the original painting by Sowerby. Sowerby contributed over 50 plates to the *Botanical Magazine*. B.L.687.c.1.

which turned out to be an unwise transaction. The astute book dealer Henry G. Bohn was the only person to benefit from the financial disaster. He bought the remaining stock and copper plates and reissued the *Flora Londinensis* in 1835 with an amended title-page. In 1847 he advertised a copy for £30 (the original price had been £87) and an elegantly bound set for £36, making the point that 'as the present reduced price scarcely repays the cost of colouring, there is no probability that the work will ever be reprinted'.

Several of William Curtis's gardening friends and members of his London Botanic Garden had expressed a desire for an authoritative work on the numerous new and exotic plants from overseas which they were trying to grow in their gardens, glasshouses and conservatories. Curtis quickly realized that here was an opportunity to recoup some of the losses incurred by the disappointing sales of his *Flora Londinensis*. In 1786 he had six plants drawn and engraved for inclusion in a horticultural magazine which he launched in February 1787. Its verbose title-page announced that the *Botanical Magazine* intended to portray 'the most ornamental foreign plants' with guidance on 'the most approved methods of culture' for the benefit of 'such Ladies, Gentlemen and Gardeners as wish to become scientifically acquainted with the plants they cultivate'. Each octavo issue would contain three hand-coloured engravings 'drawn from the living plant, and coloured as near to nature as the imperfections of colouring admit'. For many years it maintained a punctual appearance on the first day of each month. It cost a shilling an issue and from the start was a bestseller. The Revd Samuel Goodenough, in his obituary of William Curtis,[35] wrote that 3000 copies of each issue were sold; Samuel Curtis, however, stated that it enjoyed a circulation of 2000 copies during its proprietor's lifetime, but even this lower figure is impressive.

Curtis obtained plants from his London Botanic Garden and London nurserymen for his artists to draw. James Sowerby (1757–1822), the principal artist on the *Flora Londinensis*, drew the first plate (*Iris persica*) in the *Botanical Magazine*. He ceased working for Curtis in 1790 when he started a serial publication entitled *English Botany*, which his former employer viewed as a treacherous act – especially since its format resembled that of the *Botanical Magazine*, even adopting the same blue wrappers. Sydenham Teast Edwards (1768–1819), another artist on the *Flora Londinensis*, took Sowerby's place and also engraved his own drawings until about 1792, when Francis Sansom, who engraved the *Flora Londinensis* plates, took over. Etching with acid, a less laborious process than using a burin to cut a copper plate, was employed. William Graves, another member of the *Flora Londinensis* team, was chief colourist.

It was inevitable that the *Botanical Magazine*'s success would encourage rivals. Edward Donovan brought out the *Botanical Review* with 'figures of the scarcest and most beautiful foreign plants' in 1789, but after only seven numbers it ceased in 1790. *The Botanist's Repository for new and rare plants* was a much more serious contender. Its owner, Henry C. Andrews, a botanical artist and engraver, justified its appearance in an editorial comment that 'the greatest part [of the *Botanical Magazine*] … consists of those well-known common plants, long cultivated in our gardens'. Andrews, who concentrated on new introductions during the lifetime of his periodical from 1797 to 1815, had made a pertinent criticism. Curtis certainly had a predilection for the familiar flowers of the English garden, but he satisfied enough subscribers for him to boast that the *Botanical Magazine* had 'brought him pudding' whereas the *Flora Londinensis* merely offered him 'praise'.

In his will Curtis expressed a wish that his brother should continue to perform the accounting and clerical duties associated with the *Botanical Magazine* while his friend John Sims, a physician and botanist, would edit it. Sims, who renamed the publication *Curtis's Botanical Magazine*, had been left sufficient drawings to fill it for the next four or five years. A substantial increase in the cost of paper compelled him to raise the price per issue to one shilling and sixpence; by about 1808 this had soared to three shillings and sixpence and the circulation dropped to about 1000 copies per issue.

From 1811 *Curtis's Botanical Magazine* was published by Samuel Curtis, who had married William Curtis's daughter. After his wife's death in 1827 he became proprietor. His attempts to improve sales were not helped by the resentment of Sydenham Edwards,

35. *Gentleman's Magazine*, vol. 2, 1799, pp.628–29, 635–39.

whose name did not always appear on the plates he had drawn; the final insult was appending James Sowerby's name to twelve plates executed by Edwards. This unfortunate mistake, together with Samuel Curtis's feeble management, may have prompted Edwards to resign. He took with him the engraver Francis Sansom and J. B. Gawler, the author of much of the magazine's text, to found a new monthly periodical, the *Botanical Register*, with the same objectives as *Curtis's Botanical Magazine*. Messrs Weddell of Walworth, a family of botanical engravers, filled the gap left by Sansom's defection, and since there was a perceptible improvement in the hand-colouring of the plates, they probably used their own colourists. In 1818 John Curtis (no relation of the William Curtis family) became *Curtis's Botanical Magazine*'s artist.

Competition had always threatened the existence of *Curtis's Botanical Magazine*. In 1797 the *Botanist's Repository* emerged as a rival. In 1804 James Edward Smith started *Exotic Botany* because the other two could not 'keep pace with the botanical riches flowing in upon us'. Two years after the appearance of the *Botanical Register*, the nurseryman Conrad Loddiges brought out the *Botanic Cabinet*. Benjamin Maund issued his monthly *Botanic Garden* (1825–50) in two sizes: large and small paper. Robert Sweet's *British Flower Garden* struggled to survive from 1823 and only avoided extinction by merging with the *Botanical Register* in 1838. The lives of these horticultural magazines, despite their being well illustrated, was always precarious and frequently brief. By the time John Sims, now in his late seventies, retired in 1826, the circulation figures of *Curtis's Botanical Magazine* had slumped to below 1000. It desperately needed a younger man with initiative and fresh ideas to resuscitate it, to banish lethargy, to inject vitality.

The new editor came from one of the competing publications. William Jackson Hooker (1785–1865) had founded *Exotic Flora* in 1822 to illustrate 'new, rare, or otherwise interesting exotic plants, especially … such as are deserving of being cultivated in our gardens'. With a young family to support, he needed to supplement his modest salary as Professor of Botany at Glasgow. He told his father-in-law, Dawson Turner, in 1826 that provided he received £100 a year (the fee Dr Sims got) and £96 for doing the drawings, he would accept the appointment.

His editorship introduced a new phase in *Curtis's Botanical Magazine*'s history. Joseph Swan, a Glasgow engraver and printer who had illustrated Hooker's *Exotic Flora*, replaced Messrs Weddell. Commencing a new series was symbolic of the changes Hooker hoped to make. Floral dissections, which had only occasionally appeared on its plates, now became a regular feature; he made it more readable with historical and ethno-botanical anecdotes; he was eclectic in his choice of plants, giving preference to those of economic importance. Whenever particularly interesting or spectacular plants were discovered (such as the *Victoria amazonica* and *Welwitschia mirabilis*) he gave them a generous allocation of plates. As a means of avoiding the folding of large plates, he printed six copies of every issue on large paper. He pocketed the artist's fees by doing all the drawings himself – just under 700 plates. He selected plants to copy from the Glasgow Botanic Garden which he managed and from a network of gardening friends and nurserymen.

LEFT Watercolour of *Gentiana sino-ornata* by Lilian Snelling, which was published in *Curtis's Botanical Magazine* 1931, plate 9241. She contributed more than 700 plates to this magazine. Royal Botanic Gardens, Kew.

RIGHT Watercolour by William Jackson Hooker of *Gladiolus delanii* which was published in *Curtis's Botanical Magazine* 1830, plate 3032. He annotated his painting with instructions for the hand-colourist: 'Mr Graves will be so good as to be particular in the colouring of this beautiful plant.' Not far short of 700 flower paintings by Hooker appeared in *Curtis's Botanical Magazine*. Royal Botanic Gardens, Kew.

But before long he was complaining about Samuel Curtis's lack of business acumen. *Curtis's Botanical Magazine* 'is never advertised', he told Dawson Turner, '& there is no bookseller who has any particular interest in promoting the sale' of it.[36] Curtis in turn deplored Hooker's choice of 'poor miserable subjects which are selected to be the figures represented, whilst so many ornamental plants do not find access & which are now getting into common cultivation'.[37] He even said that he preferred the *Botanical Register*. Curtis wanted flamboyant flowers and at one stage Hooker feared he might take over the editorship himself or find another person. Perhaps to appease Curtis, Hooker undertook in 1833 to edit that year's issues free of charge.

New rivals kept appearing: *Floricultural Cabinet and Florist's Magazine* in 1833 (at sixpence an issue it marked the arrival of cheap gardening periodicals), *Florists' Magazine* in 1835, *Botanist* in 1836 and *Floral Cabinet and magazine of exotic botany* in 1837. A frank statement on the paper wrapper of *Curtis's Botanical Magazine* for February 1838 informed its readers 'that such a number of competitors should injure the sale of *Curtis's Botanical Magazine* was to be expected, and has, indeed, caused a considerable loss to its proprietor'. Nevertheless, Samuel Curtis vowed to continue his magazine 'with increased

36. 7 March 1830. W.J. Hooker Letters 1805–1832, f.437. Royal Botanic Gardens, Kew.
37. Quoted in Hooker's letter to Dawson Turner, 27 May 1832. W.J. Hooker letters 1805–1832, f.466. Royal Botanic Gardens, Kew.

Watercolour by Walter Hood Fitch of Fortune's double yellow rose (*Rosa odorata pseudindica*). This was published in *Curtis's Botanical Magazine* 1852, plate 4679. Royal Botanic Gardens, Kew.

vigour', promising that his 'plates shall increase in beauty, and the descriptions exceed in interest, those of any of the preceding volumes'.

This public statement was a smokescreen to hide the fact that he had been unsuccessfully negotiating its disposal. Curtis had progressively become more in debt to Sherwood, Gilbert and Piper, who had published the magazine on his behalf since 1827. By 1844 Sherwood was effectively the magazine's manager. In July 1844 Hooker informed his father-in-law that the Curtis family were in great financial difficulties. Samuel was 'without a shilling', one son lived in a cottage, and his daughter, who lived with another brother, was 'unable to fold and stitch the numbers as heretofore ... I have reason to believe that all the proceeds of the work are now made over to Sherwood'.[38] Hooker contemplated starting a similar publication entitled *Kew Gardens* or *Plants of Kew*.

In 1841 Sir William Hooker (knighted in 1836) had left Glasgow to become the first Director of the Royal Botanic Gardens, formerly part of the royal estate at Kew. He took with him a young botanical artist he had trained, Walter Hood Fitch (1817–92), to illustrate his books. His first contribution to *Curtis's Botanical Magazine* appeared in the issue for October 1834 and for many years Fitch was its sole artist.

The January 1845 issue of the magazine introduced a Third Series which exploited its editor's Kew connection. Its new subtitle promised to present 'the plants of the Royal Gardens of Kew and of other botanical establishments in Great Britain'. It displayed as its logo a wood-engraved vignette of the Palm House then under construction. In the meanwhile a purchaser for *Curtis's Botanical Magazine* had at last been found: Lovell Reeve, an aspiring young publisher who had just contracted himself to publish the *Botany of the Antarctic voyage* by Sir William Hooker's son, Joseph. The issue for July 1845 was the first under new ownership and the price remained at three shillings and sixpence. A few issues later copper engraving was replaced by lithography.

With *Curtis's Botanical Magazine* now seemingly on the threshold of a promising future, its oldest rival, the *Botanical Register*, was in terminal decline. Its merger with Sweet's *British Flower Garden* had not improved sales. Competition from cheaper horticultural periodicals had fatal consequences for quality productions like the *Botanical Register*, which in a last effort to survive proposed an alliance with *Curtis's Botanical Magazine*. Lovell Reeve dismissed this overture and the *Botanical Register* ceased publication at the end of 1847.

But all was not well with *Curtis's Botanical Magazine*. When sales fell to about 300 copies a month in mid-1848, Reeve imposed economies: a reduction of two plates in each issue and a new basis for calculating Sir William Hooker's fees.

Plagiarism was another of Reeve's problems. In 1850 he complained to the publishers of the *Gardeners' Magazine of Botany* about their unauthorized copying of *Curtis's Botanical Magazine* plates. For years the Belgian periodical *Flore des Serres et des Jardins de l'Europe* had been appropriating illustrations from *Curtis's Botanical Magazine* and other

---

38. To Dawson Turner, 31 July 1844. W.J. Hooker letters 1833–44, f.532. Royal Botanic Gardens, Kew.

English publications. Reeve had no objections, provided permission had been sought. John Gould, who published bird books, frequently adapted floral details in *Curtis's Botanical Magazine* for use in his *Monograph of the Trochilidae or family of humming-birds* (1849–61) but always acknowledged his source.

When in 1852 sales of *Curtis's Botanical Magazine* still hovered around 300 copies, Reeve insisted that Hooker's fees be reduced. If he did not accept, 'I shall probably make a total change in the contents and management of the *Botanical Magazine* agreeably with the spirit of the times'[39] – an ominous warning of his dissatisfaction with Hooker's editorship. The partial colouring of the plates, just a flower and a leaf, was part of Reeve's cost-cutting exercise.

Sir William Hooker, who disliked confrontation, enlisted the help of his son Joseph, a far more outspoken man, to protest when Reeve considered publishing another horticultural periodical. Sir William even thought of resigning and producing his own magazine, a possibility which angered Reeve, who wrote bluntly to Joseph Hooker:

> *I cannot for one moment dispute your father's right to get a new Magazine published elsewhere, but when you add that you cannot see on what principle of honour I should seek to prevent him, you must excuse me telling you that under the present circumstances it would be far from dishonourable. Without waiting for the interview which I solicited, your father determines upon starting an opposition Magazine. I shall most assuredly do what I can to prevent the success of a rival, no matter from whence it proceeds, to a publication in which I have invested property, and which it is sought to crush.*[40]

Lovell Reeve, determined to have a more popular periodical, launched the *Floral Magazine* in May 1860, devoted 'chiefly to meritorious varieties of such introduced plants only as are of popular character and likely to become established favourites in the Garden, Hothouse or Conservatory'. *Curtis's Botanical Magazine* would continue 'to represent the scientific department of Garden Botany'. The Curator of the Chelsea Physic Garden was appointed editor of this newcomer and W.H. Fitch, now freelancing, became its artist. Both editor and artist were dismissed after only sixteen issues for not satisfying Reeve's concept of popularity. The periodical staggered on until 1881. Fourteen years later the firm of Lovell Reeve was still disposing of a stock of loose plates from the *Floral Magazine* 'for screens, scrap-books, studies in flower-painting, etc.'.

In August 1865 Sir William Hooker died and the reins of editorship were automatically taken up by his son Joseph, now Director at Kew. In November of the same year Lovell Reeve died and the management of the firm passed to F.L. Soper, who would not have been blamed had he, as his first act, got rid of *Curtis's Botanical Magazine*. Instead he preserved the *status quo*, changing nothing in its format, content or price.

---

39. 26 February 1852. Lovell Reeve letters, f.152. Royal Botanic Gardens, Kew.
40. 11 May 1860. Lovell Reeve loose letters, item 323. Royal Botanic Gardens, Kew.

Stained glass window commemorating William Curtis by John Hayward in St Mary's Church, Battersea.
The wreath around the portrait incorporates a few of the flowers depicted in the *Flora Londinensis*.
On either side are the coats of arms of the Society of Apothecaries and the Linnean Society.

A dispute between Joseph Hooker and Fitch over the ownership of Fitch's original drawings, which had always been presented by the publisher to Kew Gardens, led finally to his resignation in 1876. Over a period of 40 years he had drawn about 2700 plates for *Curtis's Botanical Magazine* and his abrupt departure created a crisis. Joseph Hooker recruited his daughter, sister-in-law and cousin besides other willing volunteers as artists. Eventually he trained his cousin, Matilda Smith, who remained principal artist for the next 34 years. W. H. Fitch's nephew, John Nugent Fitch, lithographed her drawings.

Sir Joseph Hooker's relinquishment of editorship in 1904 brought to a close an association between the Hooker family and *Curtis's Botanical Magazine* which had endured for 77 years. Sir William Thiselton-Dyer, Joseph Hooker's son-in-law and now Director at Kew Gardens, became editor and within a few months quarrelled with the aged Soper. His criticism of the quality of the hand-colouring especially irritated Soper:

> *It should be borne in mind that these plates have to be drawn, printed, coloured &
> issued with descriptive letterpress at less than six pence each. We will defy anyone to
> produce better plates at the price. ... We may add that we have not received a single
> complaint from any quarter except yourself.*[41]

Soper, weary of the dispute when Thiselton-Dyer persisted in his complaints, threatened to terminate the magazine, and only Thiselton-Dyer's retirement from Kew Gardens prevented it. F.L. Soper retired and the firm was managed by one of his sons, A.L. Soper. Hand-colouring was still a contentious issue and there was a minor skirmish about the ownership of the original drawings.

In 1920, anxious to be rid of this unprofitable publication, A.L. Soper offered its copyright and goodwill to Kew Gardens for £250. Kew's parent department, the Ministry of Agriculture and Fisheries, was sympathetic but H.M. Stationery Office would not recommend its purchase while it continued at a loss. The Lovell Reeve firm terminated its connection with *Curtis's Botanical Magazine* with the last part of volume 146, published in December 1920. It was saved from extinction by a proposal of H.J. Elwes, traveller, naturalist and gardener, at a dinner at the Chelsea Flower Show in May 1921. He persuaded some of his companions to join him in buying the copyright and presenting it to the Royal Horticultural Society, which agreed to resume its publication. Kew Gardens provided editorial control in return for the retention of the original drawings. It was not an easy transition. The magazine reappeared in October 1922; although the price was substantially increased, it still failed to break even and the Royal Horticultural Society had to subsidize its production.

The vexed problem of hand-colouring persisted. The artist Lilian Snelling modified her drawings to make it easier for colourists to apply simple washes of colour, but the fact had to be faced that it had become an expensive anachronism. Furthermore, it was becoming exceedingly difficult to find people prepared to undertake routine colouring to an acceptable standard. In 1947 the *Curtis's Botanical Magazine* committee investigated the use of four-colour half-tone, photolithography and collotype. Half-tone was chosen for volume 165 (1948), signalling the end of an unrivalled tradition of hand-colouring that had lasted for 160 years. Photogravure, tried in the next volume, was superseded a few years later by photolithography.

The Bentham-Moxon trustees at Kew Gardens helped the Royal Horticultural Society by paying the artists' fees, and in 1970 the Society assigned the copyright of *Curtis's Botanical Magazine* to the Bentham-Moxon Trust. Kew Gardens continued to publish it, changing its name to *Kew Magazine* in 1984. As there were many who mourned the loss of the familiar title, it resumed the original name in 1995.

Curtis's Botanical Magazine is the oldest periodical in existence with coloured plates as its principal feature. These plates, now numbering 11,000, include examples of the work of many distinguished botanical artists. They are a unique pictorial record of plant introductions and floral fashions over two centuries. The publication has outlived all its competitors, many of which sought to emulate it. It has survived conflicts between proprietors, editors and artists, and financial crises. Every time it came close to bankruptcy it was always rescued by friends and supporters who valued it.

---

41. 8 April 1905. English letters 1901–05, ff.1458–59. Royal Botanic Gardens, Kew.

10 The Temple of Flora

Dr Thornton, a contemporary and admirer of William Curtis, applauded Curtis's vision of his *Flora Londinensis* as being 'not upon a neat, diminutive, inadequate scale, but one that was equally just, magnificent, and noble, like our Empire – one truly worthy of the British Nation.'[42] The tone of this eulogy is consistent with Thornton's ambitions for his own *magnum opus*, still being planned, which would pay homage to Linnaeus, extol British patriotism, and justly reward the author.

Robert John Thornton (*c.*1768–1837) reminds one of John Hill (1714–75). Both were compulsive authors who conceived books on a monumental scale, both were impetuous, and both failed lamentably. Thornton, the son of a successful journalist, was educated privately and went to Trinity College, Cambridge, whose Master assured his mother that he had the makings of an eloquent preacher. Thornton, however, chose medicine as a career. A family legacy enabled him to spend three years at Guy's Hospital Medical School and he took his M.B. at Cambridge in 1793. He spent four years as physician at the Marylebone Dispensary before succeeding James Edward Smith as lecturer on medical botany at the united hospitals of Guy's and St Thomas's. It would seem that medicine was to be his vocation and, assuredly, would have been his destiny had it not been for a substantial inheritance from his mother.

He now had adequate funds to indulge other dreams. An Act of Parliament passed in 1811 (to which we shall refer later) states that in 1791 Thornton

> *conceived a Plan to promote Sciences and improve the Arts of Painting and Engraving in this Country; and not doubting to receive Encouragement in the Prosecution of such Plan, did select the Science of Botany, as advanced by Linnaeus and subsequent Authors, as a source of employing the first Painters and other eminent Artists of this country, which might eventually prove a national Concern and Honour.*

This laudable ambition conveys the authentic ring of Thornton's obsession. In 1793 he issued a single-sheet prospectus, 'The first lines of botanical knowledge, being a New Illustration of the Sexual System and the Natural Orders of Linnaeus'. He expressed his confidence that

> *the elegant pencil of [his] friend Mr Sowerby, and the animated and glowing colouring of Mr Creswel, will be considered an improvement in the Arts, and [he] would be happy were the inimitable talents of Mr Bishop hereby to come into more general notice, who has executed for Bartolozzi the Frontispiece to this new attempt towards extending the knowledge of Botany.*

42. R.J. Thornton, 'Sketch of the life and writings of the late Mr William Curtis', *Lectures…by William Curtis*, vol. 3, 1805, p.16.

Though Thornton never secured the services of the botanical artist James Sowerby, his grand project was taking shape.

In the meantime his fertile mind and fluent pen were producing volumes of *The politician's creed* (1795) and *The philosophy of medicine* (1796). In January 1797 he announced the forthcoming publication of the *New illustration of the sexual system of Carolus von Linnaeus*, each part with two plates and letterpress, costing one guinea, spread over twelve to fourteen parts. By prior appointment the original pictures could be inspected at Thornton's home at 3 Bennet Street, St James's. The first part made its debut the following year in a green paper cover with a pink label. The bibliographical complexity of the work that gradually emerged indicates that Thornton had a confused concept of its structure. It eventually coalesced into three distinct sections: Linnaeus's prize dissertation on the sexes of plants, 1759; 'A full explanation of the classes, and orders, of the sexual system' with 91 botanical plates and 26 portraits; 'The Temple of Flora' with 31 plates. On 1 January 1799 he issued a title-page for the last section: 'Picturesque botanical plates of the New illustration of the sexual system'. Two of the plates ('Tulips' and 'Aloe') bear the date '1 May 1798'; the remainder were engraved between '1 May 1799' and '1 October 1805'. More than twenty parts of the entire work had been issued when Thornton opened his Linnean Gallery at 49 New Bond Street in London in 1804 to exhibit paintings he had commissioned for it.

In staging this exhibition he was 'following former examples with the utmost diffidence'. John Boydell, a prosperous print publisher, had engaged eminent artists such as Joshua Reynolds, George Romney, Benjamin West, John Opie and Henry Fuseli to produce a series of genre paintings of characters and scenes from Shakespeare's plays. In June 1789 he opened the Shakespeare Gallery in Pall Mall to display them as they were executed with the intention of engraving a selection of them. Fuseli, who had painted 40 canvases illustrating John Milton's poetry, exhibited them in a Gallery of Miltonic Sublime, also in Pall Mall, during May to July 1799. Both ventures lost money for their sponsors, but this did not deter Thornton from emulating them as he urgently needed publicity to entice subscribers for his *New illustration*.

In his Linnean Gallery with 'backgrounds expressive of the country of each flower' and 'agreeably decorated with birds in the attitude of life, butterflies, transparencies, etc', he staged his 'Botanical Exhibition', subsequently renamed 'Temple of Flora'. There hung original paintings by Henderson, Reinagle, Pether, and one by 'Sydney Edwards' (possibly Sydenham Edwards?) which had been engraved for *The Temple of Flora*. Tomkins's miniatures of Queen Charlotte and the princesses were prominently displayed, and Thornton's substantial catalogue, which went through at least four editions, reminded viewers that several of his books had been dedicated to Her Majesty. A full-length portrait of Linnaeus in Lapland dress dominated an assembly of portraits of British scientists and authors. Thornton included himself in this pantheon of distinguished men, but whereas they were each allocated just one image, he represented himself with a portrait, a miniature and a bust. He also slipped in 'with extreme diffidence' his own

Robert John Thornton *The Temple of Flora*, London 1799–1807. B.L.10.Tab.40.
LEFT A group of cultivated auriculas. Painted by Peter Henderson. Engraved by Lewis and Hapgood in
aquatint, stipple and line, 1803. The plates in this book were printed in colour and finished by hand.
RIGHT The superb lily (*Lilium superbum*). Painted by Philip Reinagle.
Engraved in mezzotint by Ward, 1799.

painting of 'A group of roses'. The catalogue carried letters of commendation from
Erasmus Darwin and the professors of botany at Cambridge and Edinburgh. It also
stated that parts of *The Temple of Flora* were on sale at twelve shillings and sixpence each.
An advertisement announced the publication of his *Empire of botany* and *The grammar
of botany*, the latter to be given free to subscribers of *The philosophy of botany*. In 1805 he
transferred the exhibition to his home at 1 Hinde Street in Manchester Square.

A notice in the *Morning Herald* for 22 January 1805 recorded that 23 numbers out
of a projected 30 had been published, but the real reason for this announcement was the
opportunity it gave Thornton to reveal that the Czar of Russia had just presented him
with a diamond ring. *The Morning Herald* for 14 June 1805 offered parts 1–25 and their
continuation at 25 shillings a part; as an inducement to subscribe it added that 'the price
of this work will be advanced when finished'. With the appearance of the thirtieth
number all that remained to be published were the frontispiece and preliminary pages.
Due to 'the interrupted communication with the Continent, from its present calamitous
state', Thornton had for immediate sale 50 copies of the complete work which he would
dispose of at the original subscription price.

In his determination to create a spectacular impression Thornton overstepped the
mark: a frontispiece, an elegant title-page, a dedication and a preface were not enough for
him. In addition to the mandatory title-page we have a subtitle-page, a contents page, a
dedication to Queen Charlotte, a page which blandly says 'select plants' (all four pages in
beautiful cursive script), then three colour plates: 'Flora dispensing her favours on the

earth', 'Esculapius, Flora, Ceres and Cupid honouring the bust of Linnaeus' and 'Cupid inspiring plants with love'. A poem by Anna Seward, 'The Swan of Lichfield', concludes this excessive fanfare.

Lengthy poems compete with the engravings for attention: egregious verse by Miss Seward, the Poet Laureate Henry James Pye, Charlotte Smith, George Shaw and others long since forgotten, extolling the beauties of flowers and deploring war. More restrained are the few extracts from Erasmus Darwin, James Thomson and William Shenstone.

Thornton's claim that 'all the most eminent British artists have been engaged' is hardly substantiated by his choice. Peter Henderson was known for his portraits and genre studies, Philip Reinagle was a landscape artist down on his luck, Abraham Pether came from a family of artists with a penchant for moonscapes. Depending on which copy of *The Temple of Flora* one consulted, Reinagle contributed fourteen and Henderson thirteen plates. Occasionally two paintings were done of the same study, although only one appears to have been used at any one time. Thornton no doubt ensured that every copy had his composition of roses.

At least thirteen engravers worked in a variety of graphic processes, mainly aquatint or mezzotint, or in combination, with stipple and line added. Usually no more than three tints were colour-printed, and the engraving touched up by hand. Their craftsmanship is superb – too good, it might be argued, for some of the paintings they interpreted.

Thornton closely supervised every stage of the book, selecting the plants for illustration and recommending suitable backgrounds. 'Each scenery is appropriate to the subject,' he wrote.

> *Thus in the Night-blowing Cereus, you have the moon playing on the dimpled water, and the turret clock points XII, the hour at which the flower is at its full expanse. In the large flowering Mimosa, first discovered in the mountains of Jamaica, you have the humming birds of that country, and one of the aborigines struck with astonishment at the peculiarities of the plant.*

His desire for a dramatic composition sometimes ignored botanical accuracy. His tropical night-blowing Cereus flourishes out of doors in the English countryside. A stapelia is rendered more sinister by the addition of a snake. His Dragon Arum is one of the most menacing plant portraits.

> *This extremely foetid plant will not admit sober description. Let us therefore personify it. She came peeping from her purple crest with mischief fraught; from her green covert projects a horrid spear of darkest jet, which she brandishes aloft: issuing from her nostrils flies a noisome vapour infecting the ambient air …*

Snowdrops, carnations and hyacinths enjoy a more pastoral setting, but they were all plates appreciated by readers thrilled by Gothic novels and enchanted by romantic verse.

Thomas Bensley, one of the best printers, saw the work through the press, using superior Whatman's wove paper. It was a handsome folio, yet sales were disappointing.

Two pages from the leaflet announcing Thornton's lottery; the first prize included the original
paintings for *The Temple of Flora*. Bodleian Library, Oxford.

Thornton calculated that he had sold about 700 copies: 6 to members of the royal family,
9 to 'foreign kings and potentates', 74 to the 'English nobility', 294 to the 'Gentry', 266
to 'Medical gentlemen', 37 to 'Florists' and 14 to 'Public bodies'. In July 1809 he dropped
the price of *The Temple of Flora* and *New illustration* to a guinea a part.

His financial problems were a spur to frantic writing, in the hope, perhaps, of producing a profitable bestseller: *Practical botany* (1808), *Botanical extracts* (1810), *Elementary
botanical plates to illustrate Botanical extracts* (1810), *Alpha botanica* (1810) and *A new
family herbal* (1810). The last of these was illustrated by Thomas Bewick, who ten years
later was still trying to get paid for his work.

Thornton decided to adopt John Boydell's solution to a similar crisis. When the
failure of his Shakespeare Gallery threatened Boydell with bankruptcy, he got an Act of
Parliament passed, in 1804, authorizing a personal lottery. By this means he hoped to
raise enough cash to remain solvent until his edition of Shakespeare came out in 1805.
His first prize was all the original paintings, lesser prizes comprised various collections of
his prints. Every purchaser of a three-guinea lottery ticket received a complimentary
Boydell print. Boydell died before the draw, but his firm recouped £66,000.

Parliament passed an Act in May 1811 (51 Geo III, cap. 103) giving Thornton the
right to hold a lottery for the disposal of his paintings, engravings and specified books.
The Botanical Lottery, renamed the Royal Botanical Lottery when the Prince Regent
became its patron, had to be held before 1 December 1813. As the money raised should

not exceed £42,000, 20,000 tickets were issued at two guineas each. With Boydell in mind, Thornton made the Grand Prize all the paintings done for the *New illustration of the sexual system of … Linnaeus* including *The Temple of Flora* and a copy of each of the several books which comprised the other prizes. Sets of *The Temple of Flora, New illustration* and *The philosophy of botany*, five volumes in all, formed the second prize. The third prize was made up of complete sets of loose engravings from *The Temple of Flora*. The fourth prize was a quarto version of *The Temple of Flora* with inferior engraving and colouring. Some illustrations were completely redrawn – the Dragon Arum had an erupting volcano and flashes of lightning added, the Night-blowing Cereus lost its clock tower, and the eggs in Thornton's bird's nest had hatched into fledglings. The fifth and sixth classes offered other publications by Thornton. A condition of the lottery was that all his copper plates had to be destroyed within a month of the draw.

Thornton once more displayed the original oil paintings, this time in the European Museum in King Street, St James's, organized his six categories of prizes for scrutiny by the appointed lottery referees, and issued a prospectus promising the draw before 4 June 1812, provided one third of all the tickets had been sold. Since sales remained sluggish it was not until March 1813 that the *London Gazette* announced that the lottery would definitely take place on 6 May. As that final date approached lottery agents were still advertising the availability of tickets, offering as an inducement an engraved portrait of the Czar to all purchasers. On the eve of the lottery the Grand Prize was once again put on public display, and Thornton gave a lecture in the Argyle Rooms on Linnaeus 'for the benefit of the sufferers in Russia'. No doubt he mentioned the diamond ring the Czar had given him. The draw duly took place the next day in Coopers Hall, Basing Street.

Like so many of Thornton's speculations, his lottery failed, and he was ruined. On the last page of *The Temple of Flora* he had apologized to subscribers for its sudden termination, revealing that he had hoped to print 70 coloured engravings. He blamed the Napoleonic wars for his misfortune:

> *The once moderately rich very justly now complain that they are exhausted through taxes laid on them to pay armed men to diffuse rapine, fire and murder over civilised Europe.*

Benjamin Silliman, an American tourist, had met Thornton in June 1805 and been shown a copy of *The Temple of Flora*. He shrewdly noted in his journal that

> *Posterity will probably wonder that a work so splendid and beautiful could ever have been executed, and still more perhaps that one so unprofitable should ever have been undertaken.*[43]

What is its reputation today? It is of no value whatever to botanists. It is a nightmare for bibliographers, as every copy seems to differ in some respect in the choice and number of plates and their various states. As a work of art it can be criticized for being ostentatious, excessively theatrical and melodramatic. Yet these are the very qualities that make

it impressive. It inspired Samuel Curtis to embark on *The Beauties of Flora* (1806–20) with flowers against a landscape backdrop. He managed only ten aquatinted and stippled plates by Thomas Baxter and Clara Maria Pope in what is now an exceedingly rare book. Like Thornton, he found such an ambitiously conceived book a challenge to produce and difficult to sell. The Australian artist Paul Jones clearly had *The Temple of Flora* in mind in his flower studies in *Flora superba* (1971), placing them in the context of their habitats.

Now dependent on his earnings as a doctor, Thornton somehow continued to publish. *Doctor Cullen's practice of physic* (1816), *Juvenile botany* (1818), *Pastorals of Virgil* (1821), *Historical readings for schools* (1822), *Greenhouse companion* (1824), *Religious use of botany* (1824) and *The Lord's prayer newly translated* (1827) were all pot boilers, but one has become as much a collector's item as *The Temple of Flora*.

Thornton was the family doctor of the artist John Linnell, who recommended his friend William Blake as one of the illustrators of a new edition in 1821 of Thornton's *Pastorals of Virgil… adapted for schools*. Blake carved seventeen small wood-engravings – the only time he used this process. As they lacked the technical dexterity that Thornton expected, he had three of them re-engraved by a journeyman engraver, and he was only prevented from replacing the rest by the protests of some artist friends. These exquisite miniatures, measuring about one and a half by three inches, are idyllic interpretations of rural life. Displayed in groups of three and four, they had to be trimmed to fit the page. Samuel Palmer and especially Edward Calvert were strongly influenced by them. Thornton apologized for their inclusion in his book 'as they display less of art than genius, and are much admired by some eminent painters'.

43. *A journal of travel in England, Holland and Scotland*, edn 2, 1812, pp.178–79.

OPPOSITE
Robert John Thornton *The Temple of Flora*, London 1799–1807. Dragon Arum (*Dracunculus vulgaris*) painted by Peter Henderson. Engraved in mezzotint and aquatint by Ward, 1801. B.L.10.Tab.40.

11 *The Birds of America*

Sacheverell Sitwell declared that 'there is nothing in the world of fine books quite like the first discovery of Audubon'. One of the largest books ever printed, its size alone commands attention and an expectation which the exuberance of the drawings and the expertise of the engraver amply fulfil.

Jean Jacques Laforest Audubon was born in what is now Haiti on 26 April 1785, the illegitimate son of a French sea captain and a servant girl. When he was three his father took him to France. In order to avoid conscription in the French army, he was sent to the United States in 1803 to manage his father's farm in Pennsylvania (his forenames were anglicized to John James). In 1805 he returned to France, where he remained for a year. In later life he claimed to have studied under the French neo-classical painter Jacques-Louis David, but his work reveals none of his influence. More likely, he received tuition from one of David's pupils. At 21 he was back in America with a French partner, Ferdinand Rozier, to help him run the family farm, which through incompetent management had to be sold in 1807. Audubon preferred to develop his skills in drawing birds in pencil, pastel and watercolour. He made a 'position board', as he called it, a framework of flexible wire to support freshly shot birds in life-like postures, quickly sketching his specimens before the colours of eyes and bill faded.

A general store in Louisville, Kentucky that he and Rozier ran was visited in March 1810 by Alexander Wilson, a bird painter soliciting subscribers for his *American ornithology* (1808–14). Perhaps this chance encounter suggested to Audubon the idea of a similar project. With his wife Lucy, whom he had married two years earlier, he started another retail store, this time in Henderson, Kentucky, subsequently trading for a year with his brother-in-law in flour, pork and lard in New Orleans before returning to Henderson to open a mill. He met Wilson again in 1812, a year before he died, and although he could not yet match his draughtsmanship, he already had a surer grasp of composition. When his business in Henderson failed in 1819, Audubon landed in a debtors' prison and was declared bankrupt. He supported his family by chalk portraiture and teaching art, which he still continued to do after he became a temporary taxidermist in Western Cincinnati in 1820.

With neither aptitude nor inclination for business, he informed Henry Clay, Speaker of the House of Representatives, in a letter of 12 August 1820, of a trip he was about to make to New Orleans. He intended to enlarge the collection of bird drawings he had amassed over the past fifteen years 'with a view to publishing them'. Taking with him a talented young thirteen-year-old student, Joseph Mason, he joined a cargo boat sailing south along the Ohio and Mississippi rivers. He collected and drew birds while his companion incorporated floral compositions. Audubon never acknowledged Mason's contribution; he even deleted his signature from drawings before despatching them to the engraver, an act of ingratitude that other assistants were also to suffer. By now Audubon had attained a fluency in mixed media: pencil, pastel, watercolour and gouache. The size of the specimen he drew dictated the size of paper he used, the largest sheets being 25 by 38 inches; he adroitly contorted particularly large birds to fit them in.

In the spring of 1824 he visited Philadelphia, the home of the American Philosophical Society, the Academy of Natural Sciences, and a number of publishers. The ninth and final volume of Alexander Wilson's *American ornithology* was being edited there and Charles Bonaparte, Napoleon's nephew, was writing a supplement to it. Instead of being welcomed, Audubon was viewed as an interloper; 'the trader naturalist', some dubbed him. His flamboyant dress, his assured manner, perceived as arrogance, and his claim to a French pedigree alienated many of the townsfolk, who had felt more comfortable with Wilson's modest mien. George Ord, the editor of Wilson's posthumous volume, dismissed Audubon's drawings as inaccurate. Alexander Lawson, Wilson's engraver and now preparing plates for Bonaparte's forthcoming book, was likewise hostile. Much more sympathetic, Charles Bonaparte urged Audubon to persevere in his quest for a publisher. Audubon might have experienced difficulty in finding an engraver with a press large enough to handle the double-elephant sized prints he had in mind. An American engraver who had recently returned from England advised him to seek an engraver there.

Acting upon this advice, Audubon sailed from New Orleans in May 1826, landing at Liverpool on 21 July. He was rather peeved when a Customs official exacted a duty of twopence on each of his several hundred watercolours. The most rewarding of several letters of introduction he had brought with him was that addressed to the Rathbone family, prominent among the city's intelligentsia. Through the Rathbones he met the banker, collector and botanist William Roscoe, who organized a successful exhibition of more than 200 of Audubon's drawings, earning him £100 in admission fees. Roscoe staged a similar exhibition in Manchester, unfortunately poorly attended. The American consul there convinced Audubon that he should sell his book by subscription. Back in Liverpool Henry G. Bohn, the London bookseller, recommended a sortie to London to meet naturalists who would introduce him to the best printers, paper-makers, engravers and colourists. He should carry out a similar exercise in Paris, Brussels and possibly Berlin, to compare costs. The first number should be accompanied by a prospectus, and Bohn advised him to consider the size of the book most carefully:

> *Remember my observations on the size of your book, and be governed by this*
> *fact, that at present productions of taste are purchased with delight, by persons*
> *who receive much company particularly, and to have your book laid on the table as*
> *a pastime, or an evening's entertainment, will be the principal use made of it, and*
> *that if it needs so much room as to crowd out other things or encumber the table,*
> *it will not be purchased by the set of people who now are the very life of the trade.*

If large public institutions only and a few noblemen purchase, instead of a thousand copies that may be sold if small, not more than a hundred will find their way out of the shops; the size must be suitable for the English market.[44]

Edinburgh repeated the enthusiastic reception Audubon had enjoyed in Liverpool. Looking a colourful embodiment of James Fenimore Cooper's American frontiersman, dressed in a wolfskin jacket, hair flowing to his shoulders, he sat for his portrait and a phrenologist made a cast of his head. The prestigious Wernerian Society of Edinburgh elected him a member. 'I am positively looked on by all the professors & many of the principal persons as very extraordinary,' he proudly informed his wife.

W. H. Lizars, at that time engaged in engraving plates for P. J. Selby's *Illustrations of British ornithology* (1818–33) and Sir William Jardine's edition of Alexander Wilson's *American ornithology*, was introduced to him in October 1826. Impressed by Audubon's drawings and undaunted by their size, he offered to make some trial proofs. Audubon, delighted with the results, immediately engaged Lizars to engrave all the plates, double-elephant size (approximately 39½ by 29½ inches, or 73 by 53 cm), for a work with five plates to each part and priced at two guineas for subscribers. He sent a draft prospectus to William Roscoe for literary embellishment. The first prospectus (it was to be frequently reissued with extracts from favourable reviews to attract more subscribers) appeared on 17 March 1827. The five plates in each part of *The Birds of America* would consist of one of the largest drawings, a medium-sized one and three smaller ones. 'In every number there will be a nondescript bird, or one not generally understood to be a native of the United States. There are upwards of 400 drawings, and it is proposed that they shall comprise three volumes, each containing about 130 plates.'

While the prospectus was being considered Audubon travelled extensively in the north, exhibiting his watercolours and some specimen engravings, talking to scientific societies, never declining any hospitality even if it meant dining late ('I cannot refuse a single invitation'). He was successful in recruiting subscribers in Liverpool and Edinburgh; a visit to a Mr Selby's house in Northumberland roped in three more; Newcastle upon Tyne and Leeds were disappointing, with only three and five respectively; York committed itself to ten, Manchester to eighteen. By mid-May 1827 he had promises of 100 subscriptions and confidently predicted he would soon double that figure, an optimistic forecast that did not take account of losses. During the course of publication of *The Birds of America* 188 subscribers were to drop out. They defaulted for a number of reasons: the uncertainty of a long-term commitment; complaints about shoddy work, especially hand-colouring; lapses in the delivery of parts; and the current economic depression. In September 1827 Audubon was forced to make another tour just to collect outstanding debts.

In May 1827 he travelled south to meet London's potential customers. One of his northern subscribers, the Countess of Morton, persuaded him to trim his long hair and

44. Maria R. Audubon *Audubon and his journals*, vol. 1, 1898, p.128.

John James Audubon *The Birds of America*, London 1827–38. Vol. 3, plate 242.
Snowy egret (*Egretta thula*). Probably painted by Audubon in March 1832 in South Carolina.
George Lehman contributed the scenic background. B.L.NL.Tab.2.

wear clothes more appropriate to the capital. He failed to gain an audience with George IV, but the monarch became a subscriber and patron of the book.

Within a month of his arrival, Lizars informed Audubon that since his colourers were on strike he would send the second part to London for colouring. The tone of another letter from Lizars a few days later intimated that he was no longer interested in the project. An entry in Audubon's journal for 30 September 1827 records the end of their business association: 'I have removed the publication of my work from Edinburgh to London, from the hands of Mr Lizars into those of Robert Havell, No. 79 Newman Street, because the difficulty of finding colorers made it come too slowly and also because I have it done better and cheaper in London.'

Robert Havell, father and son, were the proprietors of a well-established firm of engravers. When terms had been mutually agreed, Lizars was requested to forward the ten copper plates he had already engraved. Four of them were textured with aquatinting by Havell and Lizar's name was ungraciously omitted from the captions in later states of some of them. Almost the entire workforce of the Havells was devoted to the production of *The Birds of America*. It has been stated that some 50 colourists were involved. The only other important natural history book which engaged their staff at the time was Priscilla Bury's *A Selection of Hexandrian plants* (1831–34) with 51 aquatints (Audubon was one of its 79 subscribers). With this encouraging improvement in their business, the Havells made the bold decision to move to more spacious accommodation at 77 Oxford Street in 1831, calling it The Zoological Gallery.

Audubon told his wife that by employing Havell he had reduced Lizar's charges by a quarter. He paid Havell £114 for engraving, printing and colouring 100 copies of each number. Charging subscribers two guineas a number yielded a profit of just over £100, out of which Audubon had to pay administrative and travelling costs. For as long as possible he managed the business single-handed, but it soon became necessary to employ regional agents and booksellers to deal with some of the subscriptions. Whenever members of Audubon's family came to England, they, too, were involved in some of the administrative chores. Audubon had accommodation in Havell's premises, where he produced oil-painted versions of some of his watercolours. The money raised from their sale went towards expenses.

In June 1828 the partnership between Robert Havell senior and his son was dissolved and Robert Havell junior now had sole control. In January 1829 Audubon arranged for a set of existing numbers of *The Birds of America* to be coloured personally by Havell for presentation to Congress in Washington, in readiness for a campaign for subscribers in the United States.

A review by William Swainson of the first 30 engravings in *The Birds of America* appeared in J.C. Loudon's *Magazine of Natural History* in May 1828. He used the word 'genius' several times in describing the six numbers so far published, deploring the fact that they had enjoyed no publicity in booksellers' catalogues or in London's print shops, nor any recognition from periodicals save a brief reference in the *Zoological Journal*. He quibbled about the accuracy of some tinting and the inclination of a bird's bill, but these were minor criticisms in an otherwise enthusiastic review. Audubon's drawing of turtle-doves 'would secure [him] the highest meed of praise, so long as truth and nature continued the same'. In commending the author for producing 'one of the cheapest [quality books] that can be purchased', he regretted that most of the subscribers came mainly from Yorkshire, Liverpool and Manchester. Audubon naturally quoted from this favourable review in reprints of his prospectus. He also included a commendation from the French scientist Baron Cuvier: 'C'est le monument le plus magnifique qui ait encore été élevé à l'ornithologie.'

That eulogy was made at the French Academy of Sciences while Audubon was in Paris in September 1828. Cuvier, who had become a subscriber, displayed Audubon's work to members. They admired it – 'Quel ouvrage!' – but deplored the price – 'Quel prix!' One of the royal courtiers persuaded Charles X to subscribe and the botanical artist Pierre Joseph Redouté obtained the pledge of the Duchess of Orléans. Audubon's French excursion netted fourteen subscribers. His journal for 13 September records a meeting with the French engraver Dumesnil when they, rather surprisingly, discussed the printing of *The Birds of America* in France:

> he told me honestly [it] could not be published in France to be delivered in England as cheaply as if the work were done in London, and probably not so well. This has ended with me all thoughts of ever removing it from Havell's hands, unless he should discontinue the present excellent state of its execution. Copper is dearer here than in England, and good colorers much scarcer.[45]

45. Maria R. Audubon *Audubon and his journals*, vol. 1, 1898, pp.315–16.

With his wife showing some reluctance to come to England, Audubon returned to the United States in April 1829 to recruit subscribers and to draw more birds, leaving J.G. Children of the British Museum in charge of his affairs. Not a single subscriber succumbed to his exhibition at the Lyceum of Natural History in New York. In Camden in Pennsylvania he drew migrating birds while his assistant, George Lehman, added landscape details. Here and elsewhere they completed between them 42 paintings in just five months. Audubon was received by President Andrew Jackson, and the House of Representatives subscribed to a copy of his book, which the *American Journal of Science and Arts* hailed as 'the most magnificent work of its kind ever executed in any country'.[46]

He persuaded his wife Lucy to join him on his return to England in April 1830, when he was dismayed to find that during his absence the number of subscribers had dwindled and non-payment of existing subscriptions had grown, due perhaps to a rumour that he had no intention of returning. Reports of inferior work particularly annoyed him. He warned Havell that 'should I find the same complaints as I proceed from one town to another ... I must candidly tell you that I will abandon the publication'. He could be a hard taskmaster and his criticisms, often concerning hand-colouring, were nitpicking.

He increased his output of oil paintings for sale to raise desperately needed capital. He had discovered this profitable source of earnings while in Edinburgh, when he paid a young Scottish landscape painter, Joseph Bartholomew Kidd, to complete them with floral backgrounds. Never at ease with this medium, Audubon taught Kidd bird portraiture, gradually leaving their reproduction in oils to him.

The Copyright Act of 1709 had lapsed, but its clauses relating to the presentation of new books to specified libraries in England and Scotland remained in force (originally nine libraries, increased to eleven after 1801). Fortunately for Audubon, books of engravings and lithographs without any text were exempt from this expensive requirement. For this reason the absence of any text in *The Birds of America* may have been a deliberate omission.[47] But at the outset Audubon had vaguely contemplated a companion work describing the birds he had drawn, with anecdotes of his experiences. Very conscious of his own inadequate education, he approached William Swainson, the author of the appreciative review of *The Birds of America*, to write it. When Swainson demanded substantial fees and joint authorship, Audubon looked for another collaborator. In William MacGillivray (1796–1852), a young zoologist living in Edinburgh, he found an ideal assistant. He accepted a smaller fee, did not insist on co-authorship, and was conscientious and industrious. When no Edinburgh publisher would take on the *Ornithological biography*, as it was called, Audubon accepted the risk of publishing it himself. Seven hundred and fifty copies of the first volume, royal octavo in size, appeared in April 1831,

---

46. Vol. 16, 1829, pp.353–54.
47. The same reason may account for the absence of any text in J.E. Gray's *Illustrations of Indian zoology* (1830–35) and Francis Bauer's *Delineations of exotick plants ... at Kew* (1796–1803), although the latter's ostensible excuse was that the accuracy of the plates rendered any text superfluous.

John James Audubon *The Birds of America*, London 1827–38. B.L.NL.Tab.2.
LEFT Vol. 3, plate 211. Great blue heron (*Ardea herodias*).
Audubon often poised his water birds at the moment of seizing their prey.
RIGHT Vol. 1, plate 17. Carolina turtle-dove (*Zenaidura macroura*). Audubon perched these
'two gentle pairs' on a branch of flowering *Stewartia* 'emblematic of purity and chastity'.

OPPOSITE
Vol. 3, plate 251. Brown pelican (*Pelecanus occidentalis*). Audubon probably painted this
splendid male specimen while he was in Florida during 1832.

priced at one guinea for subscribers of *The Birds of America* and 25 shillings for non-subscribers. Eventually subscribers received it free of charge.

Twenty numbers of *The Birds of America* completed the first volume; parts for the second volume were under way, and Audubon needed another American trip to draw water birds scheduled for the third volume. He sailed for New York in mid-July with Florida as his destination. In Charleston he made friends with John Bachman, a Lutheran minister and naturalist who was later to collaborate with him on a book on North American quadrupeds. Audubon shot and painted birds, and his travelling companions, George Lehman and an English taxidermist, Henry Ward, drew or preserved specimens.

Audubon lost no opportunity to enrol new subscribers and sent his son Victor to London to urge Havell to print more parts for this expanding American market. Lucy Audubon irritated Havell with complaints about the shortcomings of his craftsmen, and Victor, now in England, kept up the family pressure for perfection.

PLATE. CCLI

Brown Pelican
PELECANUS FUSCUS

129

The other son, John, spent three summer months in coastal Labrador skinning specimens for his father, now on one of his last major expeditions for new birds. They moved south through the provinces of eastern Canada to New York, whence, in late September, Audubon sent Havell drawings to fill the last two parts of the second volume.

He decided to delay his departure to London until he had campaigned in towns he had not previously visited in the eastern United States. A successful tour was marred by vindictive comments from his old enemy George Ord and his English ally, the naturalist Charles Waterton. They ridiculed his scientific competence – an 'ornithological impostor', Waterton called him. Ord took pleasure in Audubon's failure to attract as many American subscribers as he would have liked: 'Did he really expect to have his monstrous book encouraged by subscriptions in the United States?'

Now nearly fifty years old ('I conceive myself growing old very fast,' he told Havell), Audubon wished to speed up the completion of *The Birds of America*. Havell, having taken on another engraver, reckoned he could increase the annual output from five to eight, maybe even ten, numbers. Volume 1 had concluded in 1830 and volume 2 in 1834; volume 3 was scheduled to be finished by 1835. Acutely depressed by the unrelenting pressures of its production and fearful that he might not be able to complete it, Audubon asked Victor to be prepared to continue it to the best of his abilities.

When he landed in Liverpool with his wife and John in May 1834, he heard that Lizars in Edinburgh was spreading malicious rumours about him. William Swainson had criticized the first two volumes of *Ornithological biography*. The stock of an American edition of this work had been burned in a Boston fire. His equipment for a future expedition was lost in a New York fire in 1835. His attempts to retrieve defecting subscribers in Manchester failed. All was not gloom, however. Engraved plates were emerging at a faster rate from Havell's establishment. The quality of those in the 54th number particularly pleased Audubon: 'You may now challenge the world of ornithological engravers without any fear,' he assured Havell.

In need of more bird studies for the fourth and concluding volume of *The Birds of America,* Audubon returned home in 1836. His greatest coup in the year's trip was the purchase of 93 bird skins collected by Thomas Nuttall and John K. Townsend in the Rocky Mountains and Columbia River, a region he had little likelihood of visiting. He used them to draw about 70 of these birds of the western United States.

After his return to London in the summer of 1837 he was faced with a difficult decision. In his original prospectus in 1827 he had promised a work of about 400 plates amounting to 80 numbers. If that figure were exceeded, he faced the possibility of the withdrawal of even the most loyal and tenacious subscribers. From 1836 he started depicting several species on the same drawing to restrict the number of plates. Nevertheless, he was convinced that it would be a grave error to omit the specimens in the Nuttall and Townsend collection. Determined to include them, he extended his work by another 35 plates, or seven numbers, making a total of 435 plates. The last plate was engraved on 16 June 1838, almost twelve years after Lizars had pulled his trial proofs.

Audubon had wanted an edition of 500 sets, but according to his son Victor only about 175 were printed.

The Birds of America was Audubon's *raison d'être*, the focus of his life to which even his patient wife and family were subordinated. One admires his determination to recruit subscribers, to find new specimens to draw and to maintain exacting standards in the engraving and colouring. According to his accounts he spent £28,910 ($115,000) on publishing the work. In March 2000 a set of the four volumes was sold at Christie's in New York for £5,500,000 (the copy had come from the Marquess of Bute's library).

Ornithological biography came to an end in May 1838 with the publication of the fifth volume. In July a *Synopsis* listing in systematic order 491 species in *The Birds of America* was published. MacGillivray was the scientific adviser and editor of the *Ornithological biography*, checking the classification and nomenclature of the birds, providing anatomical details, improving Audubon's English ('smoothing down the asperities', as Audubon put it). He deserved more than Audubon's cursory acknowledgement. Audubon was reluctant to credit the help of collaborators like Joseph Mason and George Lehman. His greatest indebtedness was to Robert Havell, who not only produced some of the finest aquatints of his day, but as a competent artist was trusted by Audubon to provide backgrounds to some of his drawings. Sometimes Havell adjusted Audubon's compositions, occasionally transferring a bird to a less crowded plate. Theirs was an uneasy relationship, with Audubon threatening to end the partnership, often finding imperfections in the plates, and deploring Havell's inefficient management. Audubon owed most to the tolerance and loyalty of his wife, who relieved him of all domestic matters to concentrate on his book. She and her husband left England for good in September 1839. Robert Havell, whose business had been largely dependent on *The Birds of America*, sold his stock and premises and emigrated to the United States, where he settled just north of New York.

The Birds of America had not brought Audubon financial security. This he achieved with the sales of a royal octavo edition of his book. A *camera lucida* produced plates to about an eighth of the size of the originals. One hundred numbers were published in Philadelphia between 1840 and 1844, each with five lithographs. With 1199 subscribers it was a bestseller; an initial print run of 300 copies expanded to over 1000.

After his father's death in 1851, Audubon's son John risked a chromolithographic version, this time double-elephant size but half the price of the original. He had planned 45 numbers, but only 105 plates had been issued by 1860 when mismanagement and the American Civil War put an end to it. Lucy, now impoverished, sold her husband's watercolours to the New York Historical Society and tried to dispose of the surviving copper plates; some had been lost in a New York fire in 1845. She sold them for scrap in 1871, but a number were saved from a smelting furnace by an alert employee who had identified them. These survivors, no more than 80, were restored and now repose in American museums and universities. Alecto Historical Editions, which had handled the plates of Banks's florilegium for the Natural History Museum in London, printed from six of these plates in 1985 for the American Museum of Natural History.

12 Flora Graeca

The finest British illustrated botanical work resulting from foreign travel is unquestionably the *Flora Graeca*. Its creator never saw it, but provided for its publication in his will. John Sibthorp succeeded Humphrey Sibthorp as Sherardian Professor of Botany at Oxford in 1784. His father, a wealthy member of the landed gentry, is remembered for having given only one lecture during the 36 years he occupied the chair of botany. Neither did his son feel constrained by academic responsibilities. He was 25 and had already spent two years travelling in Europe, financed by a Radcliffe travelling fellowship awarded to him in 1781. This gave him an annual grant of £300 for ten years, provided he travelled abroad for a total of five years. To fulfil this requirement he planned a botanical survey of the Levant and the Balkan peninsula, then part of the Ottoman Empire. This turbulent region's poor roads, inadequate transport and the xenophobia of its inhabitants effectively discouraged European travellers. Some eighteenth-century expeditions there under the aegis of the French and Danish monarchs had included naturalists, but Sibthorp's private initiative was the only significant British contribution. Sibthorp chose the eastern Mediterranean, where the flora was relatively little known although historically famous through the herbal of Dioscorides (*fl.* AD 50–70).

Only transcriptions of this Greek doctor's encyclopaedia of about 500 medicinal plants survive, the best known being the sixth-century manuscript *Codex Vindobonensis*, an outstanding example of Byzantine art, displaying 479 flower paintings. Sibthorp consulted it in the Imperial Library in Vienna, probably the first Englishman to do so. He was fortunate to be lent a rare set of unpublished engravings of these illustrations by Nikolaus Joseph von Jacquin, Professor of Botany at Vienna and Director of the university's botanical garden, which proved an invaluable source of reference.

Sibthorp's meeting with Jacquin in 1784 yielded an additional bonus – the acquaintance of a young botanical artist, Ferdinand Bauer (1760–1821). Bauer's father, who had been court painter to the Prince of Liechenstein, died when he was only two years old. He and his two brothers were encouraged to draw by their mother, and their artistic talents were developed by Nobert Boccius, prior of a convent at Feldsberg. Under his tuition they painted flowers with botanical precision for a large florilegium he was compiling. They continued their training at the Akademie der bildenden Künste in Vienna.

When Ferdinand Bauer, who had drawn and possibly engraved some of the plates in Jacquin's *Icones plantarum rariorum* (1781–95), met Sibthorp he was introduced as Jacquin's 'principal draughtsman'. Sibthorp, impressed by his work, engaged him at an annual salary of £80 to be his artist on his expedition to the eastern Mediterranean. He informed his future travelling companion, John Hawkins, that 'my Painter in each

John Sibthorp *Flora Graeca*, London 1806–40.
Vol. 1. Frontispiece with view of Mount Parnassus. Each of the ten volumes has a different frontispiece,
the first seven being designed by Ferdinand Bauer. The calligrapher was Tomkins, possibly the one who had
done similar work in *The Temple of Flora*, and the engraver was Halliwell & Co. B.L.453.h.1–10.

part of Natural History is Princeps pictorum – he joins to the Taste of the Painter the Knowledge of a Naturalist, & Animal, Plant & Fossil touched by his Hand shew the Master'.

John Hawkins (1761–1841), a Cambridge graduate and a man of independent means, was a classical scholar and a naturalist with a preference for geology. About the same age as Sibthorp, they made a very compatible pair. They met in January 1787 in Istanbul after Sibthorp and Bauer had travelled in Italy, mainly by carriage, botanizing and visiting archaeological sites. While Sibthorp praised Bauer's sketches of plants, buildings and landscape, his artist was always made acutely conscious of his own inferior social position. He made effective use of the system of colour coding to which he had been introduced by Boccius. When working out of doors many flower painters either partially coloured their rapid sketches or, like Sydney Parkinson, described the colours as accurately as they could in pencilled notes. Bauer referred to a chart depicting 250 numbered strips of colour gradations, annotating his sketches with these numbers. So adept did he become with this method that he replaced it with a chart representing nearly a thousand colour shades for his Australian voyage. The concept had been employed by Dürer and some of his contemporaries, but Bauer applied it in an assured and sophisticated manner. It has been suggested that Bauer made use of the *camera obscura* for drawing landscapes.

By the time the pair reached Istanbul, Bauer had sketched some 500 plants and Sibthorp had gathered 300 new species. Encouraged by these results, and with Hawkins about to participate, Sibthorp believed that the publication of a 'Flora Graeca' (his words) was feasible. The trio visited Cyprus, Athens and Mount Parnassus. Hawkins parted company with them at Thessaloniki in August 1787, and Sibthorp and Bauer were back in Oxford by December.

Sibthorp attended to the management of his estates, the identification of his collections, and the publication of *Flora Oxoniensis* (1794). Bauer industriously converted his sketches into polished watercolours, at the rate of about eighteen a month. Notwithstanding this impressive output, Bauer was disgruntled. That he was never to forget his subordinate status was reflected not only in Sibthorp's attitude towards him but also in his salary. His brother Francis had been appointed botanical artist to Sir Joseph Banks in 1790 with an annual salary of £300, whereas Ferdinand Bauer remained on £80. When Hawkins asked Sibthorp to consider 'offering him terms greatly superior', he was not prepared to negotiate: 'More than I pay him at present neither can I afford nor am I willing to give.'

Now a sick man, Sibthorp unwisely returned in March 1794 to Istanbul, where Hawkins joined him for excursions to Zakinthos and Morea. Bauer declined to accompany him because of dissatisfaction with his salary and failing eyesight. There is no evidence that he continued working for Sibthorp after his departure for Turkey. He had completed 966 plant drawings, some partly coloured, and some of the different title-pages for the ten projected volumes of the *Flora Graeca*. Sibthorp returned to England in September 1795, exceedingly ill, and retired to Bath, where he died on 8 February 1796.

Sibthorp bequeathed his Oxfordshire estate to the University of Oxford on condition that the rents and profits it generated (not far short of £300 a year) should first be applied to the publication of the ten folio volumes of the *Flora Graeca*, each to contain 100 engraved plates. It would be complemented by an unillustrated *Prodromus Florae Graecae*. The three executors he nominated – John Hawkins, Thomas Platt and Francis Wenman – were required to appoint a suitable editor to write a text using Sibthorp's journals, papers and specimens, and to supervise the engraving of Bauer's watercolours. With the death of Wenman in April 1796, Platt attended to business matters and Hawkins provided the organization, impetus and resolution to carry the *Flora Graeca* to publication.

It was two years, however, before Hawkins, abroad when Sibthorp died, returned to England to discuss the appointment of an editor with his co-executor. Their choice, James Edward Smith (1759–1828), a prominent botanist, the owner of Linnaeus's collections and library, President of the Linnean Society of London and author of several botanical works, manifestly had the capabilities and stamina to take on this formidable task. Besides, he endorsed Linnaeus's system of plant classification, which Sibthorp had intended to use in the *Flora Graeca*. Possibly a salary of £75 a year helped to convince Smith that he should accept the editorship.

With Smith appointed, there remained only the selection of a printer and an engraver. Richard Taylor, a Norfolk printer, was chosen, perhaps for the convenience of Smith, who lived in Norwich. Engaging an engraver was fraught with obstacles. Ferdinand Bauer was an obvious choice, but Daniel Mackenzie, currently engraving the plates for Roxburgh's *Plants of the coast of Coromandel*, was favoured by Sir Joseph Banks. Bauer was certainly interested in the commission, as Hawkins reported to Smith on 15 July 1799:

> *He [Bauer] repeated his wish to undertake both the engraving and the colouring having reason to think, he says, that he could execute the former better & cheaper than Mackenzie and wished that it might be decided by a trial.*[48]

A determining factor for the two executors was the size of the operation. Sibthorp had stipulated 1000 engravings and, with a notional print run of 50 sets, 50,000 engravings would have to be hand-coloured. In November 1799 Hawkins proposed a compromise: both artists should be engaged, with Bauer allocated the more complex plants since he had handled the original specimens from which they were drawn. Hawkins was no nearer a resolution of the impasse when he wrote once more to Smith on 23 December 1799:

> *… if Bauer objects to working in concert with Mackenzie let each engrave & colour separately his own portion of the work. I am moreover of the opinion that the immensity of the work requires us to enlist in it other artists of merit such as Sowerby for I must again observe that one man whatever may be his industry will never be able to get thro' it …*[49]

---

48. Sir J.E. Smith letters, vol. 22, f.171. Linnean Society.
49. Sir J.E. Smith letters, vol. 22, f.172. Linnean Society.

LEFT Design for the frontispiece of *Flora Graeca*, vol. 6, 1827. Watercolour and pencil by Ferdinand Bauer, 1794. The wreath of flowers for each frontispiece was selected from plants in that particular volume. Department of Plant Sciences, Oxford University.

OPPOSITE
John Sibthorp *Flora Graeca*, London 1806–40. B.L.453.h.1–10.
TOP LEFT Vol. 1, plate 40. *Iris germanica.*
TOP RIGHT Vol. 3, plate 248. *Nerium oleander.*
BOTTOM LEFT Vol. 5, plate 492. *Papaver pilosum.*
BOTTOM RIGHT Vol. 6, plate 523. *Helleborus officinalis.*

All four plates are drawn by Ferdinand Bauer and engraved by James Sowerby.

Mackenzie died early in 1800, but Bauer still refused to countenance collaboration. 'I believe,' wrote Hawkins to Smith on 13 February, 'if he cannot execute the Flora Graeca in the way he likes & on the terms he has proposed he will absolutely decline it, and then heaven knows how we shall get it done.' When Sir Joseph Banks offered Bauer £300 a year as botanical artist on Captain Flinders's impending voyage to Australia in 1801, he promptly withdrew from this inconclusive debate, thus ending his connections with the *Flora Graeca*. A satisfactory solution was found by employing James Sowerby, engraver and floral painter, who had an establishment equipped to cope with such a large project. Furthermore, he had already worked for Smith. However, it would appear from the Sowerby correspondence at the Natural History Museum in London that he was reluctant to accept the commission. After his death in 1822, his son James de Carle Sowerby continued the project, but his relations with the current editor of the *Flora Graeca*, John Lindley, were frequently strained.[50]

50. R.J. Cleevely 'Sowerbys and their publications', *Journal of Society for Bibliography of Natural History*, vol. 7, part 4, 1976, pp.353–54.

Iris germanica.

Papaver pilosum.

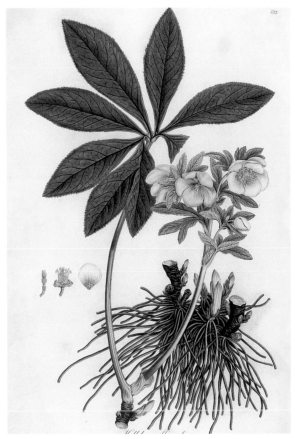

Smith was soon complaining to Hawkins about Sibthorp's illegible handwriting and the absence of annotations on his collections and drawings. Hawkins's reply that 'Dr Sibthorp did not mark all his specimens, or the drawings; but he trusted to his memory, and dreamed not of dying' was hardly a consoling explanation. Smith persevered and in February 1805 a prospectus announced the forthcoming publication of the *Flora Graeca* in 20 parts, each containing 50 engraved plates, cumulating in ten folio volumes. The first part would be priced at twelve guineas but subsequent ones might be less, dependent upon a subsidy from Sibthorp's legacy. The executors had decided to publish each volume in two parts in order to maintain a cash flow and to reduce the length of time subscribers had to wait for the next instalment. The prospectus included a brief reference to the *Prodromus Florae Graecae*, the first volume of which appeared in October 1806, marginally in front of the first volume of the *Flora Graeca*. The *Prodromus*, completed in two volumes in 1816, described 2600 species, whereas the *Flora Graeca* was confined to the 966 species drawn by Ferdinand Bauer.

With a work of such magnitude it is perhaps not surprising that Smith and the two executors were sometimes at loggerheads. Hawkins resented Smith's criticisms of his Greek orthography and his queries on the habitats of the plants he had collected with Sibthorp. He reminded Smith

> *of the high responsibility we have already incurred by advancing you £750 before a sheet of either work [Flora Graeca or Prodromus] was printed & before I believe any portion of the work was ready for the press except what is now printed but will leave you to imagine how unpleasant our situation would be in case of your demise under these circumstances.*[51]

Evidently Thomas Platt, the other executor, joined in the attack. Smith defended himself robustly:

> *But I wish you and all the world to know that the protracted publication of the work (except what rose from my being almost blind for some months, and therefore unable to work at it) has been owing to the confused nature of the state in which our deceased friend left the materials, and which no one could have suspected beforehand.*[52]

Hawkins, loyal to the memory of his dead friend, insisted on carefully checking Sowerby's engraving and colouring, constantly alert to any departure from the high standards Sibthorp would have wanted. Before his departure to Australia, Ferdinand Bauer had completed seven of the ten title-page landscapes. It was probably Hawkins who commissioned William Westall to draw the three remaining views: Mount Athos, 'Physcus' (i.e. Marmaris) and Delphi.

Smith, who was writing the text for Sowerby's *English botany* (1790–1814), seems to have considered sympathetically Sowerby's desire to distribute the *Flora Graeca*. So,

---

51. 28 December 1806 / 3–6 January 1807, Letters of Sir J.E. Smith, vol. 22, f.179. Linnean Society.
52. 1 January 1807, Lady P. Smith *Memoir and correspondence of … Sir J.E. Smith*, vol. 1, 1832, p.559.

confident that this would be permitted, Sowerby rented premises to house the stock. Hawkins firmly resisted the proposal 'because a booksellers shop is the usual & the most direct channel of communication between the Editor of a work & the Public'. He proposed John White of Fleet Street as a more appropriate agent, having 'the largest stock of books of any person in the trade besides dealing particularly in works of Natural History'. From the third volume of the *Flora Graeca*, sales were handled by Payne and Foss, booksellers in Pall Mall.

Hawkins parried Smith's quibbles while Platt struggled with the finances. Funds for printing the *Flora Graeca* came from the trustees, and subscribers were disappointingly few in number. No economies in its production were possible, since Sibthorp had stipulated a lavishly illustrated work to do justice to Bauer's drawings. It was Platt's responsibility to balance the accounts. A complete set of the *Flora Graeca* would cost about £620 to produce[53] with subscribers contributing only £254. The shortfall had to come from Sibthorp's estate. Platt needed all his cunning to counter a claim from the trustees of the British Museum for a free copy of the work.

For such an expensive publication the trustees were prepared to take legal action to obtain the free copy to which they believed they were entitled. The defendants, Payne and Foss, represented by Platt, argued that the *Flora Graeca* was not being published for profit, that it was heavily subsidized by Sibthorp's estate, and that only 30 sets were to be published for which there were 26 subscribers, leaving only a few copies for sale. It was not 'a book which legislature intended should be entered at Stationers' Hall'. The case was heard in the Court of King's Bench in January 1827 and in the Exchequer Chamber in Error a year later.[54] The judge decided that a part of a work 'published at intervals of several years, at an expense exceeding the sum to be obtained by the price of the copies, and which expense was defrayed by a testamentary donation, was holden not to be a book demandable by the British Museum under [Act of Parliament] 54G.3.c.156'. Had the British Museum won its case, ten other libraries designated by the Copyright Act could also have claimed a free copy, and had that happened the *Flora Graeca* might have ceased publication. Sir James Edward Smith (he had been knighted in 1814) died shortly after the Court's judgement, having seen through the press six volumes of the *Flora Graeca*.

Robert Brown, Keeper of Sir Joseph Banks's botanical collections at the British Museum, reluctantly agreed to edit the first part of the seventh volume which Smith had left unfinished. Very little had to be done to it, but in June 1830 when Hawkins made enquiries about its progress Brown had still not handed it to the printer. It was eventually published in June 1831, more than three years after the preceding part. Perhaps smitten by a guilty conscience, Brown gave his fee to Smith's widow. A brilliant botanist but a notorious procrastinator, Brown was obviously an unsuitable editor for the *Flora Graeca*.

---

53. In 1810 Smith told Roscoe that 'the colouring of each individual copy costs 4 shillings a plate'. Roscoe letters 2378. Liverpool City Archives.
54. P. Bingham *Report of cases argued and determined in the Court of Common Pleas*, vol. 4, 1828, pp.540–49.

A replacement was found in John Lindley (1799–1865), Professor of Botany at University College London, Assistant Secretary of the Horticultural Society of London, the author of four books, and, unlike Brown, young and energetic. He tried to impose a strict timetable on James de Carle Sowerby, several times threatening to transfer to another engraver if he persisted in falling behind schedule. In November 1840 the last part of the *Flora Graeca* appeared. Within two years both executors were dead, having endured its protracted publication for more than 40 years. They had fulfilled all Sibthorp's wishes except in the number of plates (966 instead of 1000). Their friend had envisaged a work that would favourably compare with Curtis's *Flora Londinensis*, Jacquin's *Flora Austriaca* and Oeder's *Flora Danica*. It is one of the finest books in botanical iconography, the culmination of the golden age of British botanical art stretching from the mid-eighteenth century.

Hawkins and Platt had hoped for 50 subscribers and 30 had been recruited during Smith's editorship, but only 25 remained when the tenth and final volume was printed. Being such a rarity, it is not surprising that few periodicals reviewed it. Sir William Jackson Hooker suggested another printing from the original copper plates held by the University of Oxford. H.G. Bohn, the bookseller and one of the subscribers, offered his services. By using surplus letterpress and coloured engravings and the original copper plates, he made up 40 sets which can be distinguished by the '1845' or '1847' watermarks. He advertised this reprint in his 1847 catalogue at the bargain price of £63!

Marble memorial to John Sibthorp by John Flaxman, in Bath Abbey.
Sibthorp is depicted as a traveller in Greek costume, possibly returning to England, with a symbolic posy of flowers. The building is the Danby Gate at the Oxford Botanic Garden.

13 Flora Danica

Ambitious and costly projects like the *Flora Graeca* inevitably had a long gestation period: *Plants of the coast of Coromandel* struggled on for 25 years, *Flora Brasiliensis* took 75 years, but the *Flora Danica* holds the record at 122 years. Advances in the classification of plants facilitated the compilation of national and regional floras such as Jacquin's *Flora Austriaca* (1773–78), Waldstein and Kitaibel's *Descriptiones et icones plantarum rariorum Hungariae* (1802–12) and Ledebour's *Icones plantarum … floram Rossicam* (1829–34). Had his *Flora Londinensis* (1775–98) paid its way, Curtis had hoped to extend its scope to the whole country. One of the best known of all national floras is the *Flora Danica*, partly because of its associations with a magnificent porcelain service.

In 1735, during the reign of Frederik V of Denmark, a National College for Commerce and Economy was established to revitalize the country's moribund industry and trade. An agricultural depression demanded a programme for cultivating derelict land and developing forestry. A promotion of the natural sciences in the university curriculum was seen as a preliminary step towards agricultural reform. A medical student went to Uppsala to study under Linnaeus and a young German from Bavaria, Georg Christian Oeder (*c.*1728–91), an economist and statistician as well as a physician and botanist, was invited to apply for a teaching post at the University of Copenhagen. When some of the xenophobic staff thwarted this appointment of a German by failing his public defence of a thesis given in Latin (a necessary procedure for any professorship), the King, who took an interest in gardens and arboriculture, commanded him to plan a botanical garden and a botanical institute in Copenhagen. With a salary from the Privy Purse, Oeder during 1752–53 prepared a report which recommended a demonstration garden, a library, public lectures, and the training of head gardeners and foresters. Government officials in Denmark's colonies were to report on their natural resources. Oeder himself would travel throughout the King's domains, studying plants and their uses.

When the King approved his plans for the garden and library, Oeder, now enjoying professorial status, visited botanical gardens in Europe, hoping at the same time to find an artist and engraver to work on a publication he had proposed. He envisaged it embracing an introduction to botany, a list of Danish plants in a pocket-book format, copper engravings, and an account of the uses of plants. At that time Norway was linked with Denmark and so the book's coverage was to embrace much of Scandinavia as well as Iceland, Greenland, the Faroe Islands, Schleswig and Holstein, and Oldenburg and Delmenhorst in northern Germany. Oeder expected the engravings to be issued as loose items, enabling purchasers to arrange them as they pleased.

During his European travels he had found three candidates for the post of artist and engraver. Each one submitted a curriculum vitae and samples of his work. Christian

Eberhardt claimed he could do eight drawings a month, but had no engraving skills. Georg W. Baurenfeind, who had worked for Trew in Nuremberg, could likewise turn out eight drawings a month and also engrave them. Oeder preferred P.J. Kaltenhofer, who had illustrated the anatomical works of Albrecht Haller, despite his admission that his monthly output would be only six drawings; but he, too, was an engraver. The controller of the Privy Purse, the Lord High Steward, submitted the candidates' drawings to the staff at the Royal Sculpture and Painting Academy for their opinion and asked for any nominations from amongst their pupils. These adjudicators agreed that Kaltenhofer's work was superior, but were equally unanimous in rejecting his proposed salary as too generous for the lesser skills required of a flower painter. Rather than approve a German candidate, they recommended one of their own students, Jens Lund. Although not a flower painter, he confidently claimed that he would acquire the necessary skills to satisfy Oeder within a few weeks. One of the Academy's staff would fulfil the duties of engraver. This proposal was submitted to the royal physician and the royal surgeon for consideration. They observed that since Oeder was about to go collecting in Norway and Jens Lund would be expected to accompany him, if appointed he would have had insufficient time to be trained. With the elimination of Lund, Kaltenhofer was offered the post, but he unexpectedly withdrew. It appears that he had used his application to negotiate an increase in his salary at Göttingen!

Oeder was now instructed to find a botanical artist during his tour of England, the Netherlands, France and Germany. Georg D. Ehret, a German artist establishing a successful career for himself in England, was not interested. In Nuremberg Oeder signed a contract in 1755 with Michael Rössler (1705–77), an engraver of several of Haller's books, appointing him as his engraver, with his son Martin Rössler (1727–82) as botanical artist.

Oeder and his newly appointed botanical artist returned from Norway at the beginning of 1760. In a prospectus Oeder circulated inviting subscriptions for a book to be entitled *Flora Danica*, it was announced that the text would be published separately from the engravings, which would be issued in batches of 60 to each part. The first part came out in 1761 in plain and coloured versions, but Oeder had too few colourists working for him to satisfy the demand for the coloured plates. With a grant from the Privy Purse he employed a skilled colourist, Mrs J. A. Seizberg, to colour those engravings destined for the King. He now had orders for 185 copies of the first part and had yet to receive requests from abroad. He calculated that an expert colourist could handle annually 20 to 22 sets, each of 60 plates. As he was unhappy about the quality of the work of his colourists (who, for their part, were dissatisfied with their rates of pay), he sent 45 sets for colouring to Brunswick, where not only could his brother supervise the work but where salaries were lower. Oeder's desperation is apparent from his suggestion that Norwegian subscribers might find their own colourists. In February 1762 he discussed with the Royal Sculpture and Painting Academy the possibility of forming a school to train specifically women as colourists, since the academicians had rated hand-colouring as too menial a task for men. In April Mrs Seizberg and Mrs Starck (who had coloured Rösel's book on insects) were appointed to run an 'Illumination School' for 'Womenfolk' until sufficient colourists, recruited from a local orphanage, had been trained.

Mainly Norwegian plants were figured in the first two parts of the *Flora Danica*. Johann G. König, who was later to go to the Danish settlement in India, where he would botanize with Roxburgh, collected plants in Iceland. With a backlog waiting to be drawn, Oeder engaged Kaltenhofer in Göttingen to draw plants common to both Göttingen and Copenhagen in readiness for a future part. Since the Danes resented having their flora drawn abroad, he brought Johann Christoph Bayer (1738–1812) to Copenhagen to assist hard-pressed Martin Rössler. Part ten in 1771 was the last to appear under Oeder's editorship. He had supervised the production of 600 plates, an achievement overlooked by his Danish enemies, who wanted him banished from Copenhagen. His connections with a fellow German, Johann Friedrich Struensee, an influential politician in the Danish court who was executed for adulterous relations with the Queen, provided them with an opportunity. In 1772 Oeder was exiled to German-speaking Oldenberg, where he joined the judicial service. The school for women colourists was closed and the *Flora Danica* temporarily ceased publication.

It resumed publication in 1775 under the editorship of Otto Friedrich Müller, whom Oeder had proposed as his successor. Johann Theodor Bayer (1782–1873) – the son of Johann Christoph Bayer, who had become a painter with the Copenhagen porcelain factory in 1776 – contributed some 1500 plates to the *Flora Danica* until 1867. From about 1850, possibly earlier, Johann Christian Thornam (1822–1908), a pupil of J.T. Bayer, drew for the *Flora Danica*, until its demise in 1883. After Müller's death in 1784 a succession of eight editors followed him in the production of one of the world's greatest national floras, with 3240 engraved and lithographed plates in 51 parts and three supplements consolidated in eighteen volumes.

However, it never fulfilled its objectives for the Danish Government. The most useful section, on the commercial uses of plants, was never published, nor was the official distribution of parts satisfactorily carried out. Oeder had expected the clergy to play a pivotal role in making the *Flora Danica* available. Each diocese and county received two plain copies for consultation by interested citizens, who, in turn, were expected to pass on their observations to the botanical institute in Copenhagen. Unfortunately, the distribution of copies was inefficient, sets were incomplete, and those that were received were stored in chests and presumably forgotten. This unsatisfactory method of distribution was changed and copies were subsequently deposited in more accessible diocesan libraries.

National pride as well as public utility had motivated the launching of the *Flora Danica*. Oeder saw it as a model which other countries in Europe might emulate, their joint efforts resulting in an illustrated European flora. The only botanist to subscribe to this vision was Jacquin, who published his *Flora Austriaca* as a supplement to the *Flora Danica*. William Curtis rejected a proposal that he should omit from his *Flora Londinensis* plants that had already been featured in the *Flora Danica*.

G.C. Oeder and others *Flora Danica*, Copenhagen 1781–1883. Royal Botanic Gardens, Kew.
TOP LEFT Vol. 1, plate 138. *Utricularia vulgaris.* TOP RIGHT Vol. 1, plate 170. *Plantago uniflora.*
BOTTOM LEFT Vol. 2, plate 194. *Primula veris.* BOTTOM RIGHT Vol. 2, plate 314. *Scabiosa columbaria.*

144

LEFT AND CENTRE Two items from the Flora Danica porcelain dinner service, decorated with illustrations from volumes of the *Flora Danica*. De Danske Kongers Kronologiske Samlang.

RIGHT Wedgwood 'Water Lily' service. This pattern is composed of foliage and flowers selected from *Curtis's Botanical Magazine* and H.C. Andrews's *Botanist's Repository*. British Museum, London.

The decoration of a royal porcelain service of more than 2000 pieces with flowers taken from the *Flora Danica* may have been conceived by Theodor Holmskjold (1732–93), a former student of Linnaeus, a Privy Councillor, Lord Chamberlain of the royal household, Professor of Natural History and Medicine, and Director of Copenhagen's botanical garden before managing the Danish Porcelain Factory. No archives which positively confirm the origins of this celebrated porcelain ware have survived. The first mention in the records, dated 28 August 1790, instructed the royal librarian to send a copy of the *Flora Danica* (then consisting of seventeen parts with 1020 engravings) to the Danish Porcelain Factory. No one has been able to confirm or deny the long-standing belief that it was to be a gift from Christian VII of Denmark to Catherine the Great of Russia. Such a gesture might have been influenced by Josiah Wedgwood's table and dessert service of 952 pieces, decorated with English landscapes and country houses, made for the Empress. Another hypothesis is that Catherine had ordered the set and on her death in 1796 Czar Paul had cancelled the commission, whereupon the Danish monarch ordered its continuance. B.L. Grandjean, a former archivist at the Royal Copenhagen Porcelain Manufactory, favours the suggestion that Frederik, the Crown Prince, had intended presenting the botanical service to Catherine, but has no idea why it was abandoned.

By September 1792, 988 pieces had been made; two years later the service had expanded to over 1400 pieces. It had originally been intended for 80 guests, a figure increased to 100 in 1797. Over the next few years the rate of production declined and in 1802 the Crown Prince ordered its cessation – much to the relief of the Factory's directors, now losing money on the project. The King's birthday on 29 January 1803 was the first occasion on which the service was used. Storing it was always a problem. The Lord

Chamberlain found temporary accommodation for it in the Chinese room at Rosenborg Castle. Whenever it was brought out for use some pieces, usually plates, were broken. At a royal dinner in 1841 a record number of 32 plates were smashed or damaged. Frederik VII gave away pieces as presents; 43 items were lost in a fire at Christiansborg Castle in October 1884. In 1906 a keeper of the collection was found to have sold 101 pieces.

Johann Christoph Bayer, who became the Factory's first decorator in 1776, painted the service until he retired for reasons of poor eyesight in 1801. As far as he was able, he allocated large plants from the *Flora Danica* to large pieces of the service and diminutive plants like fungi and mosses to smaller pieces, in order to avoid distortion or truncation in copying the engravings. Whenever in doubt about the details of any engraving he would consult living specimens in the Botanical Garden. He painted about 1800 pieces and his assistant Christian Nicolai Faxoe (1763–1810) decorated 158 items.

After an interval of nearly 60 years the manufacture of the porcelain service was resumed in 1862. Old ceramic moulds which had survived the bombardment of Copenhagen in 1807 were repaired to make 725 pieces for 60 people as a wedding present for Princess Alexandra, daughter of Christian IX, who was betrothed to the Prince of Wales, later Edward VII. Living specimens from the Botanical Garden as well as illustrations from the *Flora Danica* decorated it. Chromatic colours replaced the former sombre tints. Royal Copenhagen Porcelain still adds a few pieces to the service; 1530 pieces of the 1802 set are kept at Rosenborg Castle.

The Meissen factory in Germany, which discovered how to make Chinese porcelain during the early eighteenth century, at first embellished its ware with Chinese chrysanthemums and peonies before substituting European flowers. Other factories imitated Meissen before finding their own distinctive styles. Flowers were treated naturalistically by Bow, Chelsea, Coalport, Derby, Rockingham, Spode, Swansea and Worcester. Derby pioneered the practice of adding the names of plants on the base of its pieces. Chelsea was the first manufacturer in England to copy or adapt flower illustrations in published works.

A floral decoration used by Chelsea during its Red Anchor period in the mid-1750s was for many years generically described as 'Sir Hans Sloane's Plants'. Research carried out by Dr Bellamy Gardner during the 1930s showed this appellation to be misleading. The plants had been copied from Philip Miller's *Figures of the most beautiful, useful and uncommon plants* (1755–60). Miller worked at the Chelsea Physic Garden, which paid a peppercorn rent to Sir Hans Sloane, Lord of the Manor of Chelsea, so the link of this Chelsea ware to Sloane is a tenuous one. Some of Chelsea's designs were accurate copies of parts of flowers in Miller's book, others were composite arrangements of foliage and flowers from different engravings, the aim of the porcelain painter being to achieve a balanced composition. Miller had employed five artists, of which the most competent was Georg D. Ehret, his sister-in-law's husband. Chelsea's selection of Ehret drawings came from the first volume of Miller's book, C.J. Trew's *Plantae selectae* (1750–73) and Ehret's own book, *Plantae et papiliones rariores* (1748–59). The scattering of butterflies on 'Sir Hans Sloane's Plants' plates were probably also lifted from insect drawings by Ehret. In addition, Chelsea appropriated illustrations in George Edwards's *Natural history of uncommon birds* (1743–51).

The Derby factory plundered the first nine volumes of *Curtis's Botanical Magazine*. When John Sims became editor in 1801 it pleased him to draw attention to the fact that 'a large proportion of the ornaments of our most expensive porcelain and cabinet ware' was copied from his magazine. An extensive porcelain service made in Berlin, 1817–23, relied on plates in *Curtis's Botanical Magazine* and H.C. Andrews's *Botanist's Repository* (1797–1815). Derby likewise found suitable illustrations in Andrews, including his *Coloured engravings of heaths* (1802–30). James Sowerby's *English botany* was another source, and especial favourites were the charming compositions in John Edwards's *A collection of flowers* (1795).

No one was surprised that the painters at the Cambrian Pottery in Swansea were required to copy flower engravings, since its proprietor, Lewis Weston Dillwyn, was also a botanist. Disliking free, impressionistic plant portraits, he insisted on his painters studying the accurate flowers in the first fourteen volumes of *Curtis's Botanical Magazine*. In order to ensure verisimilitude, Dillwyn had copperplate copies made of the engravings in Curtis and impressions from these copper plates were transferred to his dishes, plates and pots. It was then a routine matter to fill in the outlines by hand. He used these 'botanical patterns' on earthenware made between 1812 and 1817, not only achieving mirror images of *Curtis's Botanical Magazine* engravings but also operating a cheaper process with semi-skilled staff.

The Wedgwood Factory also delved into *Curtis's Botanical Magazine* and the *Botanist's Repository* to decorate a botanical dinner service which has since been variously called the 'Nelumbium', the 'Lotus' or the 'Water Lily' service. It was once believed to have been made for Erasmus Darwin and so earned the sobriquet 'Erasmus Darwin' or 'Darwin' service. Here is another instance of misattribution. Josiah Wedgwood was supposed to have made it for his friend Erasmus Darwin. Sir Joseph Hooker, once the owner of a set, stated that it had been produced to celebrate the publication of Erasmus Darwin's *The botanic garden* (1789). The series of 150 pieces for 24 people came into production in late 1806. Its design incorporated Sydenham Edwards's drawing of *Nelumbo* published in *Curtis's Botanical Magazine* in February 1806. Since this is after the deaths of both Josiah Wedgwood and Erasmus Darwin, it could not have been made for the latter. There is evidence, however, that it might have been purchased by Robert Darwin, Erasmus's son and Charles Darwin's father, hence the putative Darwin association.

14 Nature printing and related methods

The application of a coloured substance to the surface of a reasonably flat object, which is then pressed onto a sheet of paper, is a delightfully quick way to obtain a facsimile copy of it. Children have played with this basic technique, decorators and scientists have refined it. The earliest surviving example is preserved in the papers dated 1425 of Conrad von Butzenbach in the Studien Bibliothek in Salzburg. Here a plant had been coated with a green pigment to obtain a crude silhouette. Evidence of another early attempt can be found in the documents of Leonardo da Vinci, who applied lamp-black mixed with oil to a sage leaf. Alexis Pedemontanus in his *Liber de secretis naturae* (1557), the first book to describe a method of nature printing, recommended flattening fresh leaves with a wooden pestle to crush any thick veins. His distorted impressions were intended as decorative motifs and not as a botanical record. Botanists who used nature printing preferred dried specimens, with their greater firmness and rigidity. Thomas Horsfield, who botanized in the East Indies between 1802 and 1818, used nature prints as a quick way of recording his collections.

By the late seventeenth century lamp-black and rubbing by hand had been replaced by printer's ink and a flat press. J. Peele's *The art of drawing and painting in water-colours* (1731) described a 'way of printing the leaves of plants so that the impression shall appear as black as if it had been done in a printing press'. Particularly innovative was his 'method of taking-off the leaves of plants in Plaister of Paris, so that they may afterwards be cast in any metal'. Johannes Hieronymus Kniphof (1704–63), librarian at Erfurt's Academy of Naturalists and from 1737 a professor at Erfurt University, entered into partnership in 1728 with J. M. Funcke, a local printer and bookseller, to publish a series of nature prints using a flat press and printer's ink. In this venture they were encouraged by local nursery-man Christian Reichart. The 200 plates of *Botanica in originali pharmaceutica* (1733) were a precursor to a more ambitious work in twelve parts, each with 100 plates, entitled *Botanica in originali seu herbarium vivum…* (1757–64). The title-page of each part had an attractive floral border. Nature-printed foliage provided the structure for most of the plates, while fragile flowers and fleshy berries were added by hand. Not many copies were printed as the process was laborious and specimens would have to be frequently changed. Experiments with nature printing have revealed that as few as 10 impressions can be taken before a plant becomes limp and misshapen through over-inking, although as many as 30 can be obtained with small plants of a simple shape. Other German publishers followed Kniphof's lead: J. J. Hecker's *Specimen Florae Berolinensis* (1742) and J. M. Seligmann's *Die Nahrungs-Gefaesse in den Blattern* (1748) with skeleton leaves printed in red. The clarity of their impressions make the 800 plates in D.A. Hoppe's *Ectypa plantarum Ratisbonensium* (1787–93) among the best of eighteenth-century nature prints.

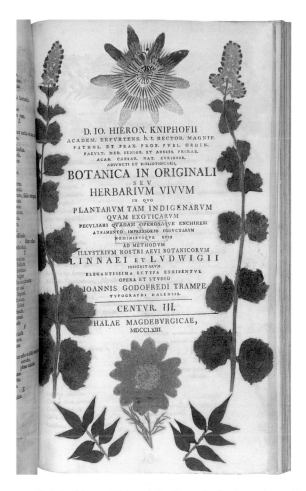

Johann Hieronymus Kniphof *Botanici in originali, seu herbarium vivum*, Erfurt 1757–64.
Title-page of one of the twelve centuries, or gatherings, of 100 plates.
Nature printed with printer's ink and a flat press. B.L.448.K.12.

In eighteenth-century Britain nature printing stagnated, still treated merely as a curiosity. William Gilpin, with his wife's assistance, perpetuated the primitive method of using the sooty deposit from the flame of a candle for leaf impressions in his book *On the variety of the forms, and the internal tracery of leaves* (1794). R. Dosie's *Handmaid to the arts* (vol. 1, 1796) reminded practitioners 'to bruise the leaves, so as to take off the projections of the larger ribs, which might prevent the other parts from plying to the paper'. In North America Benjamin Franklin ingeniously used casts of leaves to print on paper currency in an effort to outwit counterfeiters. In the nineteenth century it was lace rather than plants that stimulated an interest in nature printing in Britain. Lace manufacturers in Nottingham produced their pattern books in the 1840s by several versions of the nature printing technique. One required the lace to be placed between two blocks of boxwood, the surface of which had been softened by steaming, to take an impression of the lace. The blocks were dried, the lace removed and the block with its intaglio design inked to print a white silhouette of the lace.

In 1833 a Danish goldsmith and engraver, Peter Khyl, tried out a technique whereby the contours of leaves, feathers, lace, etc., sandwiched between a sheet of hard polished steel and one of soft lead, could be impressed onto the softer sheet. Despite experimenting with copper, zinc and tin as well as lead, he never succeeded in translating this promising theory into practice, and his manuscript on the subject lay forgotten in the archives of the Imperial Academy in Copenhagen until 1853. Nevertheless, he had discovered how to transfer impressions to metal plates by employing steel rollers. The next stage in the process, the use of electrotyping, was introduced by Dr Ferguson Branson, physician to the Sheffield Infirmary. He had made tolerable impressions of ferns in gutta-percha, but the subsequent prints were invariably framed by a dirty margin. Realizing that only a smooth metal surface, easily wiped clean, would prevent such blemishes, he recommended electrotyping impressions. However, the trouble and cost involved deterred him from pursuing this solution. He also experimented with brass casts from gutta-percha moulds of ferns and tested the suitability of Britannia metal as a medium for making impressions. But British investigations never got beyond an exploratory stage.

It was the Imperial Printing Office in Vienna which seized the initiative and discovered how to make nature printing commercially viable. Professor Franz Leydolt had employed the resources of the Imperial Printing Office to reproduce the surfaces of flat minerals by etching and gutta-percha impressions. These efforts were displayed at the Great Exhibition of 1851 in London's Hyde Park. The first nature print executed by Alois Auer, the Imperial Printing Office's Director, was a fossilized fish from an electro-typed gutta-percha cast. When gutta-percha presented technical difficulties Auer's assistant, Andrew Worring, substituted soft lead. Auer, who described this process as 'Naturselbst-druck' (natural self-printing), published an account of it in 1853; he reprinted it in 1854, accompanied by more than 50 examples, and, with an international audience in mind, supplied a text in German, English, French and Italian. He added details of his patent to defend his claim of being its inventor. Auer may not have been pleased when the Director of the royal collections of engravings in Copenhagen announced the discovery of Khyl's manuscript, which 20 years earlier had proposed the use of steel and lead plates, a key component of his own invention.

Predictably, Auer had been outraged when the English printers Bradbury and Evans took out a patent in June 1853 for 'Improvements in taking impressions and producing printing surfaces (communicated to us from abroad)'. The communicator was Henry Bradbury, the proprietor's son, who had inspected the Viennese establishment a year earlier. Auer responded in print in January 1854, denouncing the *Conduct of a young Englishman Henry Bradbury at the Imperial Government Printing Office in Vienna*. He had graciously consented to Bradbury's written request to witness

> *a certain process of printing facsimiles of dried flowers ... [in order to] introduce it into our establishment in London. In being then permitted to introduce it, my greatest pleasure and study would be, to fully acknowledge the said permission in every way.*

Auer's resentment was provoked when the Bradbury and Evans patent failed to acknowledge its indebtedness and when Henry Bradbury retorted, somewhat tactlessly, that in any case the English patent related to improvements on Auer's method. *The Athenaeum*

RIGHT
David Heinrich Hoppe
Ectypa plantarum Ratisbonensium,
Regensburg 1787–93.
Vol. 1 plate 3. *Angelica sylvestris*. A good specimen
of nature printing using printer's ink. B.L.36.g.13.

BELOW
Thomas Moore and John Lindley
The ferns of Great Britain and Ireland, London.
Nature printed by Bradbury and Evans, 1855–56.
Royal Botanical Gardens Kew.

LEFT Plate 1. 'The common Polypodium'
(*Polypodium vulgare*).
RIGHT Plate 45. 'Common Maidenhair fern'
(*Adiantum capillus-verneris*).

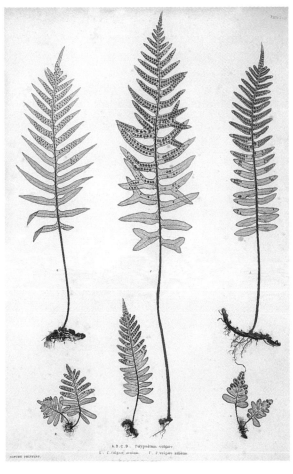

for December 1853 patriotically supported the home side in this dispute:

> *Sufficient has been here said to prove that priority of invention and application is of English origin and that the patent under which the patent must be worked in this country, is an English one.*

In his lecture on nature printing given at the Royal Institution on 11 May 1855, Henry Bradbury continued the dispute. He reminded his audience that Peter Khyl had thought of the combination of steel and lead plates and the use of steel rollers, and that Ferguson Branson had contributed electrotyping. Auer, backed by the resources of a government department, had exploited this earlier research. His lecture, entitled *Nature-printing, its origin and objects*, was published in 1856, dedicated, perhaps cynically, to Auer.

When Bradbury gave this lecture his firm had already published *A few leaves represented by 'nature printing' showing the application of the art for the reproduction of botanical and other natural objects with a delicacy of detail and truthfulness to nature unattainable by any other known method of printing* (1854). It consisted of 21 plates, without any text, and sold for a guinea, with individual plates priced at one shilling and sixpence each. It was clearly a publicity exercise to demonstrate the potential of the process.

It became a curtain-raiser for Thomas Moore and John Lindley's *Ferns of Great Britain and Ireland*, published in seventeen green-wrapper parts, each six shillings, or six guineas for the leather-bound volume. In this folio a few of the 51 prints show the marks of the burin and punches used to clarify the structural details of the ferns. The intaglio lines of the printing plate itself had been inked first with the darkest colour (usually the roots), then with a sequence of progressively lighter colours. According to Bradbury, the printer 'is not only able by this means to blend one colour into another, but to print all the colours at one single impression'. John Lindley, the botanical editor, applauded the result in his preface:

> *It is true that Nature Printing has its defects as well as its advantages; it can only represent what lies on the surface, and not the whole even of that. But, on the other hand, its accuracy is perfect as far as it goes; and in the case of British Ferns it goes far enough for all practical purposes.*

Printed under great pressure, these prints have what Bradley called a 'raised or embossed appearance', and by lightly passing one's finger over them the contours of stems and leaves can be felt.

Bradbury's next book, the four volumes of W. G. Johnstone and A. Croall's *Nature-printed British seaweeds* (1859–60), has 219 delicately coloured plates, many of which have artistic appeal as charming designs. A two-volume *Octavo nature-printed ferns*, with 122 plates of smaller fern specimens, was published about the same time. Bradbury had planned further volumes in this series on exotic ferns, British mosses and lichens, and native and exotic trees, but after his suicide on 1 September 1860, aged 31, Bradbury and Evans abandoned the process.

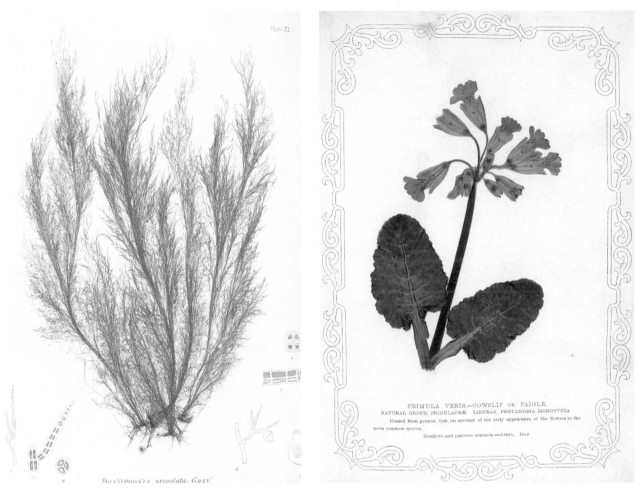

LEFT William Grosart Johnstone and Alexander Croall *Nature-printed British seaweeds*, London 1859–60. Plate 11. *Polysiphonia urceolata*. Nature printed by Bradbury and Evans. B.L.7033.Add.13.

RIGHT May Howard *Wild flowers and their teachings*, Bath 1848. Cowslip (*Primula veris*), one of a number of dried plants stuck in the book. B.L.C27.K.11.

A few individuals remained loyal to it, or to variations of it. F. O. Hutchinson in his *Ferns of Sidmouth* (1862) placed an ink-coated fern on a lithographic stone and printed the resultant image. For some inexplicable reason nature printing enjoyed a brief vogue in southern India. Readers of the *Madras Athenaeum* in 1848 were instructed in 'easy and expeditious methods of taking facsimiles of plants'. The Madras Government Press produced *Specimens of nature printing from unprepared plants* (1857), and the Basel Mission Press at Mangalore published a limited run of the Revd J. Hunzikar's *Nature's self printing* (1862), with excellent coloured plates. The outstanding achievement of this briefly fashionable graphic process is the *Physiotypia plantarum Austriacarum* (1855–73) by Constantin von Ettingshausen and Alois Pokorny, published by the Imperial Printing Office in Vienna in ten volumes.

A nature print, especially if it is as heavily embossed as the ferns in Moore and Lindley, seems incredibly real and infinitely preferable to the two-dimensional lithograph in Hutchinson. The shape of some inflorescences eliminates them as candidates for nature printing; the most suitable are non-fleshy ones such as ferns and algae. Since nature print-ing destroys diagnostic details such as plant hairs and fleshy tissue, it can only serve the botanist as an *aide-mémoire*. The great strides made in photo-mechanical processes killed any possibility of its survival in publishing.

It is a short step from taking impressions of plants to illustrating books with the actual plants themselves. For years botanists had been in the habit of compiling a *hortus siccus*, or a collection of dried plants glued or sewn onto paper sheets and assembled into the format of a book. In the sixteenth century Hieronymus Harder made a living out of producing and selling these. Eventually this well-established practice evolved into sets of such books complete with a printed title-page and contents page and descriptive labels. Grasses in particular lend themselves to drying and pressing, and an early example is George Swayne's *Gramina pascua* (1790), a selection of common pasture grasses gathered together on six sheets. The London nurseryman James Dickson issued *A collection of dried plants* (1789–91) in four parts, and during 1793 to 1802 nineteen parts of *Hortus siccus Britannicus*, dedicated to William Curtis. Curtis was himself urged by one of his clients, T.G. Cullum, to publish sets of dried grasses: 'I am ready to subscribe my five guineas if you will undertake the scheme.' Curtis must have heeded this advice, because in 1802 he was selling a *Hortus siccus gramineus* in two large volumes. Its title-pages were printed, but the plant names were handwritten. Much more professional was *Hortus gramineus Woburnensis* (1816) by George Sinclair, gardener to the Duke of Bedford at Woburn, with specimens of over 320 different species and varieties of grasses. When it was reprinted in 1824, lithographs replaced the dried specimens. W.J. Hooker praised Edward Hobson's *Musci botanici; a collection of specimens of British mosses and hepeticae* (1818–22), commenting 'how much superior these collections must be, in point of accuracy, to the best of plates'.

During the 1840s the Bath firm of Binns and Goodwin enchanted the public with reasonably priced gift books displaying pressed plants within decorative borders 'suitable for the Library of the Connoisseurs, the Study of the Agriculturists, and the Drawing Room of the Affluent'. They included F. Hanham's *Natural illustrations of the British grasses* (c.1846) and Mary M. Howard's *Wild flowers and their teachings* (1845) and *Ocean flowers and their teachings* (1847).

The disadvantages of this form of illustration are obvious. It is a laborious task to find, collect and press suitable specimens of which there may be a shortage. Since no specimen is absolutely identical with another, it cannot be cited in the way one cites a conventional illustration. Editions are of necessity small; the contents are fragile, bulky and easily damaged or destroyed by insects.

At the beginning of the Victorian era traditional graphic art processes were about to be challenged by the invention of photography. A pioneer woman photographer, Anna Atkins, was also a botanist. This was a time when social conventions confined women's creativity to genteel occupations such as flower painting ('It is a most quiet, unpretend-ing, womanly employment,' the novelist Mary Mitford enthused) and the collecting and pressing of plants. Elizabeth Fulhame, author of *An essay on combustion* (1794) and one of Sir John Herschel's research workers, strongly resented the attitude of

> *some [who] are so ignorant that they grow sullen and silent and are chilled with horror at the sight of anything that bears the semblance of learning, in whatever shape it may appear; and should the spectre appear in the shape of a woman, the pangs which they suffer are truly dismal.*

Some women had an affinity with marine life, exploring rock pools and collecting algae. Mrs Amelia Griffiths enterprisingly dried the specimens she found on the Devon coast and neatly mounted them in albums for sale. Her former servant, Mary Wyatt, followed her example and produced *Algae Danmoniensis* (1835), embellished with 200 dried seaweeds. Mrs Margaret Gatty haunted the shoreline and gave an account of her discoveries in *British sea-weeds* (1862). R.K. Greville gratefully acknowledged such female contributions to algology in his *Algae Britannicae* (1830):

> *To them we are indebted for much of what we know on the subject … the very beauty and delicacy of the objects have ever attracted their attention.*

Anna Atkins (1799–1871), although a seaweed enthusiast, had other interests as well. She drew and painted in watercolours and learned how to lithograph. She meticulously drew more than 250 shells for her father's translation of Jean Lamarck's *Genera of shells* (1823). Her father, John George Children, was a Fellow of the Royal Society and also its Secretary. He had entered the British Museum as an assistant librarian in the Department of Antiquities, eventually becoming the first Keeper of its Zoological Department. He had arranged for the American bird paintings of Audubon to be shown at the Royal and Linnean Societies and was one of the proposers for his membership of the Royal Society. During Audubon's absence from Britain he occasionally kept an eye on his business affairs.

On 21 February 1839 Children chaired a Royal Society meeting at which W.H. Fox Talbot gave members a résumé of his experiments with 'photogenic drawings'. In January 1839 Louis Daguerre had told Parisians of his invention of the daguerrotype, which, using a camera, produced an image on a sensitized copper plate. It was this news that had prompted Fox Talbot to disclose to his London audience his own photographic researches. His method scored over Daguerre's by enabling multiple copies to be produced. He made superfine writing paper light-sensitive by soaking it in solutions of common salt and silver nitrate. He first tried copying leaves, flowers and lace by placing them on this treated paper and giving them a prolonged exposure to the sun. The negative image of the object was 'fixed' by a strong salt solution.

In thanking Fox Talbot for sending him some examples of these 'photogenic drawings', Children wrote that 'my daughter & I shall set to work in good earnest 'till we completely succeed in practising your invaluable process'. At this stage Sir John Herschel becomes a participant in these formative years of photography. It was he who established the use of the word 'photography', who defined the terms 'negative' and 'positive' in this context, who produced an effective 'hypo' or fixing agent, and who invented the cyanotype process. He sent a copy of his paper on cyanotypes which he had read to the Royal Society in 1842 to Children. The process was inexpensive, easy to operate, and unlike Fox Talbot's early attempts it guaranteed a permanent image. It involved brushing plain paper with a solution of ferric ammonium citrate and potassium ferricyanide and leaving it to dry in a dark place. When an object was placed on the sensitized paper under a sheet of glass, to keep it in contact with the paper, and exposed to sunlight, a faint image would appear within five to fifteen minutes. Washing in clean water rendered the white image permanent against the Prussian blue background colour that is so characteristic of this process. The result was, to be precise, a photogram rather than a photograph, since no camera and negative were involved. For many years draughtsmen and engineers have used these 'blueprints' for duplicating their drawings and plans.

In October 1843 the first part of *Photographs of British algae. Cyanotype impressions*, in a blue wrapper, was presented to a few privileged friends by its author, Anna Atkins. She was to spend the next ten years copying further seaweeds from her own and friends' collections. Each of her illustrations was obtained by a separate exposure, so during multiple copying the positioning of the specimen on the page must have varied slightly each time. The text – title-page, contents, introduction and captions – consisted of cyanotype copies of her handwriting on transparent paper. In her preface she apologized for a few indistinct images due to the thickness of some specimens, which made it impossible to press the glass 'sufficiently close to them to ensure a perfect representation of every part'. This project had scientific objectives, but her artistic sensibilities must have been excited by the ghostly white images: fortuitous patterns suspended in an appropriately blue ground.

With the conclusion of the third volume in September 1853 and the production of 389 plates and fourteen pages of text, her task was done. The thirteen surviving copies (or fragments) do not share the same number of plates. The British Library copy, for example, has 411 plates, whereas the one in the Art Gallery and Museum at Glasgow has 439. It was Atkins's practice to provide a replacement plate whenever she found a superior specimen to copy; in at least one instance she replaced a whole part. As the recipients seldom discarded their replaced plates, variation in the number of plates in each set has occurred.

It is not known how many copies of the work were produced, but the labour involved in its production suggests that it would have been few. Possibly three parts of *British algae* had already been sent out before Fox Talbot published the first number of his *Pencil of nature*, containing photographic images, in June 1844. Thus the distinction of being the first book to be published with photographs belongs to *British algae* – a bibliographical landmark. It is, however, of limited use to algologists as the morphological details are imprecise and no information is given about collecting localities.

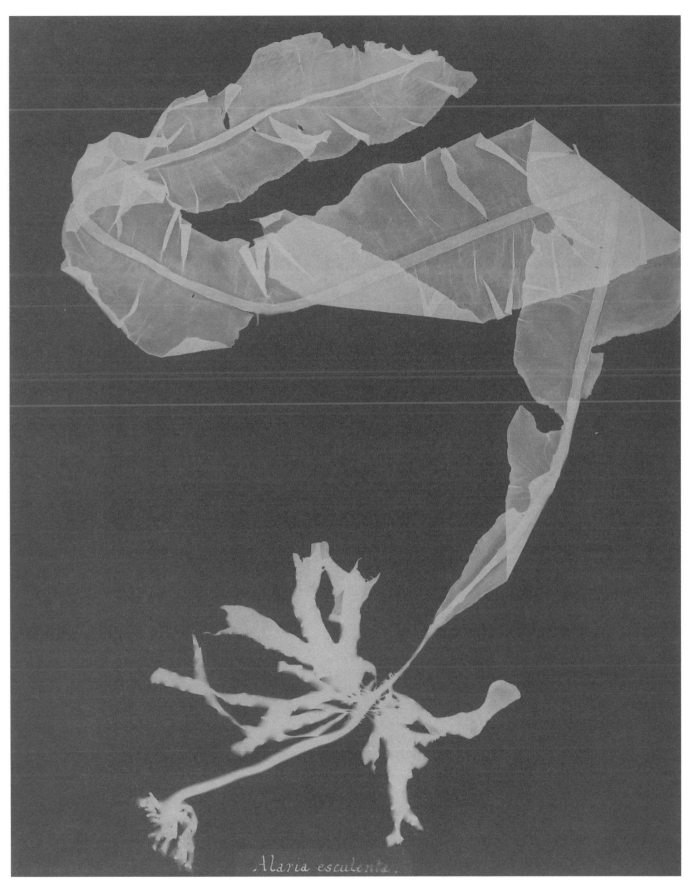

Anna Atkins *Photographs of British algae. Cyanotype impressions*, 1843–53.
B.L.C192.c.1.

Strictly speaking, Anna Atkins was not the first person to make use of photography in a publication, as Gavin Bridson reminds us.[55] George William Francis sensitized the surface of boxwood blocks, on which he placed some plants. These were exposed to light to produce photograms, which were then cut out by the wood engraver. The *Magazine of science and school of arts* (27 April 1839, 25) published examples of this ingenious method. H.G. Heath combined nature printing and photography in the fourth edition of his *Fern paradise* (1878). 'The fronds here represented were laid upon white card-board ... By a slow and laborious method their actual impress was taken, and the identical figures thus obtained were photographed on the blocks of the engraver.'

55. *Printmaking in the service of botany*, 1986, p.141.

Anna Atkins *Photographs of British algae. Cyanotype impressions*, 1843–53.
B.L.C192.c.1.

15 Clerics as authors

The Church has always presented Nature as a manifestation of God's omnipotence and munificence. Some in holy orders have lectured on aspects of it in schools and universities or written about it in books and magazines. J.C. Loudon, horticultural writer and founder of the *Magazine of Natural History*, advised the country parson to make it his recreation:

> *It is superior, in a social point of view, even to a taste in gardening … the naturalist is abroad in the fields, investigating the habits and searching out the habitats of birds, insects, or plants, not only invigorating his health, but affording ample opportunity for frequent intercourse with his parishioners. In this way their reciprocal acquaintance is cultivated, and the clergyman at last becomes an adviser and friend, as well as a spiritual leader.*[56]

Constricted by isolation or deprived of congenial neighbours, many rural clerics turned to natural history or antiquarian studies for intellectual stimulation. There has probably never been an English county without a parson diligently observing the changing seasons. Eminent naturalists such as John Ray (who had himself taken holy orders) were indebted to them for local records.

It was not until Elizabethan times that the parson, now a graduate and often from the landed gentry, showed any scholarly inclinations. The Revd Charles Butler (*c*.1560–1647), vicar of Wootton St Lawrence in Hampshire for 47 years, wrote *The feminine monarchie or treatise concerning bees and the due ordering of them* (1609). Two centuries later another entomologist, the Revd William Kirby (1759–1850), collected over 150 specimens of bees in his parish for his *Monographia apium Angliae* (1802). But few of these early naturalists aspired to publication. The Revd Adam Buddle (*c*.1660–1705), who collected grasses and mosses for much of his life, drafted a new English flora which never got beyond the manuscript stage.

The most scientific clerical author of his day was the Revd Stephen Hales (1677–1761), perpetual curate at Teddington in Middlesex, with livings in Hampshire and Somerset. His introduction to Newtonian physiology and chemistry at Cambridge was a preparation for *Vegetable Staticks* (1727), the outcome of his experiments in plant physiology. Although a Fellow of the Royal Society and enjoying the friendship of the Prince and Princess of Wales, he never neglected his parochial duties. Indeed, his scientific work only strengthened his religious belief: 'The further researches we make into the admirable scene of things, the more beauty and harmony we see in them: and the

---

56. D.E. Allen *The naturalist in Britain*, 1976, pp.22–23.

159

Gilbert White *The natural history of Selborne*, London 1789. Frontispiece.
'North-East view of Selborne from the Short Lythe.' November 1788. 'At the foot of this hill … lies
the village which consists of one single straggling street, three quarters of a mile in length, in a sheltered
vale, and running parallel with The Hanger [i.e. the beech wood.].' G. White. B.L.G2432.

stronger and clearer conviction they give us, of the being, power and wisdom of the divine Architect.'

In Georgian England a spacious vicarage in its own grounds proclaimed a comfortable living. The Revd James Woodforde (1740–1803) lived and, as his diary confirms, dined well on £400 a year in his Norfolk parish. A similar stipend was enjoyed by another *bon viveur*, the Revd Sydney Smith (1771–1845), who stoically endured his exile from society in a remote Yorkshire parish, and denounced the country as 'a kind of healthy grave'. Some fellow sufferers escaped boredom by joining the hunting, shooting and fishing fraternity. Self-sufficient individuals like the Revd Charles Kingsley (1819–75) rejoiced in 'the monotonous life of a country parson … It is pleasant and good to see the same trees year after year; the same birds coming back in spring to the same shrubs; the same banks covered with the same flowers.'

Such uneventful contentment, regulated by seasonal rhythms, appealed to the best known of all parson-naturalists, Gilbert White (1720–93). He was born and lived much of his life in the small Hampshire village of Selborne, where his paternal grandfather had been vicar. By choosing to take his degree at Oriel College, Oxford, he became ineligible for the living at Selborne, which was in the gift of Magdalen College. The fellowship awarded him by Oriel College and the curacies at Faringdon and Moreton Pinkney gave him few duties and a modest income supplemented by the rents on properties in Selborne inherited through his mother. When the incumbent at Selborne was briefly ill, he became curate-in-charge. In 1763 the family home, The Wakes, passed to him. A bachelor of frugal habits, he could afford to remain in Selborne.

The village, dominated by The Hanger, a beech-clad hill, with coppices, hedgerows and a water meadow, provided a rich ecological setting for plants, small mammals, birds and insects. It became his outdoor laboratory, subjected to his inquisitive eye and critical attention. Weather patterns and the effect of climate on plants, bird migration, the hibernation of swallows, the behaviour of crickets – his interests were omnifarious. When he was observed shouting at bees through a large trumpet in order to test their hearing, or searching at night by candlelight for earthworms in his lawn, no one objected. By the mid-eighteenth century clerics no longer wore sober black in conformity with a solemn mien. The parishioners' spiritual leader could now dress casually and even carry a sporting gun without comment.

White filled notebooks with his observations. In his 'Garden Kalendar', started in 1751, he recorded crops he had grown and their yields. In his 'Flora Selborniensis' he listed wild and cultivated flowers he had found, alongside incidental comment on birds, insects and the weather for the whole of 1766. He also noted the local flora in his copy of William Hudson's *Flora Anglica* (1762).

White probably met the zoologist Thomas Pennant in the London bookshop of his brother Benjamin White in 1767. At that time Pennant was engaged in collecting natural history data in specific localities and White's activities at Selborne interested him. From this chance meeting began an exchange of letters on topics of mutual interest. In 1768 the Hon. Daines Barrington, lawyer, antiquarian and naturalist, presented White with a copy of his *Naturalist's journals* which he had printed in 1767. It consisted of pages with ruled columns to be completed by the recipient. A week was allocated to each page for daily observations on temperature, wind, rainfall, weather, 'trees first in leaf', 'plants first in flower', 'bird and insects first [to] appear or disappear' and miscellaneous comments. Its use was intended to ensure a comprehensive and consistent coverage which Barrington hoped might eventually serve as the basis of a general natural history of Great Britain. As it appealed to White's sense of order and method, he began using it straight away, and Barrington joined Pennant as the second of his two principal correspondents on natural history matters. They urged White to consider consolidating his data into a book. He told his brother that he might write up the natural history of Selborne, covering one calendar year. Barrington persuaded him to be more ambitious, to embrace a broader perspective:

> *In obedience to yr repeated injunctions I have begun to throw my thoughts into a little order, that I may reduce them into the form of an annuus historico-naturalis comprising the nat. history of my native place. As I never dreamed 'til very lately of composing any thing of this sort for the public inspection, I enter on this business with great diffidence, suspecting that my observations will be deemed too minute & trifling. However if I ever finish it I shall submit it to yr better judgment.*[57]

Over the next few years this concept of a book took shape. White informed his brother on 29 April 1774 that he proposed extracting from his correspondence over the past 50 years enough material to 'make up a moderate volume' on the natural history of Selborne.

57. White to Barrington, 12 February 1771. Add MS 31852. British Library.

To these might be added some circumstances of the country, its most curious plants, its few antiquities, all of which together might be moulded into a work, had I the resolution and spirits enough to set about it.[58]

He sought his brother's assistance in sorting the letters and arranging the journals, promising: 'I will publish.' The fulfilment of his resolve had to wait nearly fifteen years. White had plausible excuses for the delay. In March 1775 it was 'stiffness in my eyes from overmuch reading'; in May 1776 'his left hand is full of gout'; in November 1780 he complained that 'much writing and transcribing hurts me'. Still he persevered. In 1776 he had commissioned a Swiss artist, Hieronymus Grimm, to draw twelve views of his village for the book. In January 1788 he had reached the final stage, the compilation of the index.

Although the *Natural history and antiquities of Selborne* is dated 1789, copies were available from late the previous year. Its title-page announced no author, his identity only being revealed at the end of the 'Advertisement' or preface. It is a collection of letters to Pennant between August 1762 and November 1780, followed by 66 letters to Barrington from June 1769 to October 1781. Another 26 letters on Selborne's antiquities conclude the quarto volume. A number of the letters were written specifically for the book and had never been posted to either correspondent. Some of the genuine letters were truncated, others fragmented and rearranged to conform to a theme. The book was not an immediate success; the second edition had to wait until 1802 and the third until 1813. Its popularity seems to have started in the 1830s, since when it has never, or seldom, been out of print, with more than 200 editions to its credit.

Gilbert White was a diffident author, but neither the Revd George Crabbe nor the Revd Charles Kingsley was similarly inhibited. It has been said that George Crabbe (1754–1832) took holy orders to obtain an income while he wrote and published poetry. He was also drawn to geology, botany and entomology, as well as poetry. He wrote on the fauna, flora and fossils of the Vale of Bevoir and the plants of Framlington and contributed to D. Turner and L.W. Dillwyn's *Botanist's guide through England and Wales* (1805).

Charles Kingsley (1819–75) was also a poet but is better remembered as a novelist. With the rectory at Eversley in Hampshire as his refuge, he played a more active role than Crabbe in the popularization of natural history. He was confident of a time to come when 'young men [would join] natural history societies; going out in company on pleasant evenings to search together after the hidden treasures of God's world'. In 1871 he founded the Chester Society of Natural Science, Literature and Art. In the same year he told students at Sion College that he sometimes dreamt of

a day when it will be considered necessary that every candidate for ordination should be required to have passed creditably in at least one branch of physical sciences, if it be only to teach him the method of sound scientific thought.

58. G. White *Natural History and Antiquities of Selborne*, edited by T. Bell, 1877, vol. 2, 28.

He had a younger audience in mind when he wrote *Glaucus or the wonders of the sea-shore* (1855). It had been serialized in the *North Briton Review* and went into several editions.

Kingsley, like so many of his fellow clerics, was curious about the origin of fossils. *Reliquiae diluvianae* (1826) by the Revd William Buckland (1784–1856) presented a plausible thesis for all diluvialists. A belief that geological evidence confirmed the description in Genesis of the Great Flood was still stubbornly supported in 1888 by the Revd W.B. Galloway (1811–1903) in his *Science and geology in relation to the universal deluge*. There can be little doubt that nineteenth-century clergymen, anxious to reconcile religious belief with scientific theory, quite fortuitously lent a powerful impetus to geological research.

Birds were favourite candidates for study. The Revd F.C.R. Jourdain, a notorious egg collector and author of *Eggs of European birds* (1906), later renounced his predatory habits when he became an apostle of conservation. The Revd Churchill Babington (1821–89), also a compulsive birdwatcher, earned the nickname 'Beetles' because of his affection for these insects. The Revd Octavius Pickard-Cambridge (1828–1917) is honoured as the 'father of British spiders'. The Revd William Kirby (1759–1850) co-authored the *Introduction to entomology* (1815–26), a popular manual in four volumes with amateurs.

The Revd William Turner (*c*.1508–68) has been called the 'father of English botany'. Nowhere is an indebtedness to the clergy more evident than in the 36 volumes of *English botany* (1790–1814) by Sir James Edward Smith and James Sowerby. Smith frequently acknowledged either specimens they sent him or their discovery of new localities.

During parochial visits a parson might indulge in casual botanizing. From time to time he might report his observations to the journal of his local natural history society. If a county flora were in the course of compilation, he would very likely be a contributor. The published floras of the following counties were all edited by clergymen: Bedford (1798), Oxford (1833), Shropshire (1841), Hertford (1849), Sussex (1887), Hereford (1889), Suffolk (1889), Kent (1899), Derby (1903), Norfolk (1914), Devon (1939), County Durham (1988). These county floras were primarily for use by botanists and knowledgeable amateurs and not intended for the general public seeking information on plants at a superficial level. There were, however, parsons who could write popular and semi-scientific books in an engaging style.

The ordinary reader appreciated the anecdotal text of the *Natural history of British birds* (1850–57) by the Revd Francis Orpen Morris (1810–93). The Revd Charles A. Johns (*c*.1818–75) wrote some ornithological primers: *British birds and their haunts* (1862) and *Bird life and the Bible* (1887). *His Flowers of the field* (1851) had reached its 35th edition by 1925 and was still in print in the 1970s. The popularity of his small-format books may have been helped by the fact that they were published by the Society for the Propagation of Christian Knowledge.

The most prolific popularizer of natural history was the Revd John George Wood (1827–89) with more than 40 titles to his credit. He had briefly tried pastoral work before finding writing more congenial. The demand for his books, especially from young readers, kept them in print for many years. His *Common objects of the country* (1857) sold 100,000 copies in the first week of publication, a triumph not matched until the Revd Keble Martin's bestseller just over a century later.

William Keble Martin (1871–1969), like his father and grandfather, chose the Church as his vocation. Educated at Marlborough College and Christchurch, Oxford, where he graduated in philosophy, botany and church history, he served as a curate in three parishes in the north of England before becoming vicar of Wath-on-Dearne in Yorkshire. Twelve years later he was transferred to Devon, where he was appointed rector of Haccombe, Great Torrington and Combe-in-Teignhead successively. This peripatetic career encouraged rather than hindered his passion for the British flora. During his residence in Devon he co-edited the *Flora of Devon* (1939).

When he was about 22 years old he drew a few snowdrops backed by ivy leaves, the very first drawing – as he declared in his autobiography *Over the hills* (1968) – for a book that was to make him famous. At the time it was just an idle sketch, not the start of a project, but fortunately it was his habit to keep his drawings. At first he drew flowers at random, only adopting a more systematic approach after obtaining a copy of H.C. Watson's *London catalogue of British plants*, which 'was marked off carefully into 100 sections with usually 15 to 20 names in each section'. He allocated a large sheet of paper to each section, which he gradually filled with drawings of related species. This method, quite unintentionally, presaged the layout of Keble Martin's future book. Whatever time could be spared from parochial duties was given to painting flowers. After Sunday services he would frequently take a late train to visit a promising site, sometimes as far away as Scotland, usually drawing the specimens he had collected on the return journey. Having drawn about 700 plants by the mid-1930s, almost half of his tentative total, he resolved to complete the 100 sheets, perhaps now thinking of eventual publication. He redrew those of his earlier paintings which were badly discoloured; others were replaced when better specimens were found. He travelled tirelessly, even getting to the small island of Steepholme in the Bristol Channel to find the peony reputedly brought there by monks in the Middle Ages. On another excursion he collected a rare saxifrage on Ben Lawes, drew it, and returned next day to replant it. Other botanists sent him specimens which he might have had difficulty in collecting himself. All his drawings, each on a separate sheet with its name, location, and date of collection, are notable for their accuracy and clarity. They were the components of his rather tight floral compositions on the larger sheets of paper.

After a successful exhibition of some of his drawings at the Royal Horticultural Society in January 1959, the search for a publisher began. Seven publishers in turn rejected it as too costly to reproduce in colour. An appeal for funds was launched in 1961 and His Royal Highness The Duke of Edinburgh, who received a copy of the appeal letter, asked to see a sample of Keble Martin's drawings. In 1964 the book packager George Rainbird took a gamble and offered to market the book, provided a lower standard in colour fidelity was acceptable. In May 1965 *The concise British flora in colour*, published by Ebury Press and Michael Joseph, at last came out. The Duke of Edinburgh's foreword undoubtedly helped sales, but the story of an author now 88 years old, who had laboured

William Keble Martin *The concise British flora in colour*, London 1965. Plate 57. Primulaceae. B.L.X311/271.

William Keble Martin *The concise British flora in colour*, London 1965. Plate 83. Iridaceae. B.L.X311/271.

for 60 years in his book's compilation, also appealed to the public. The drawings were attractive and the price very reasonable. Over 100,000 copies had been sold by 2 December, when the *Daily Express* declared it to be 'one of the publishing phenomena of this or any other decade'. By 1972 about a quarter of a million copies had been purchased. The grateful publisher, Michael Joseph, published Keble Martin's autobiography in 1968 and the Postmaster-General invited him to design a set of four wild flower stamps.

Since Keble Martin does not mention gardening in his autobiography, one wonders if he was that rare creature, a country parson who derived little enjoyment from cultivated plants. The large plot of land that often surrounded a vicarage usually tempted its occupant to grow flowers and vegetables. The Revd John Laurence (1668–1732) advocated gardening for 'clergymen and other studious persons that have a taste for beauty and order'. When he was not gardening he was writing bestsellers. His first book, *The clergyman's recreation: shewing the pleasure and profit of the art of gardening* (1714), had reached its sixth edition by 1726. Laurence's preface submits his apologia for this indulgence:

> *If anyone shall now say, upon sight of this little treatise that, as a clergyman, I might have employ'd my time much better than to write about gardening; I answer, that it*

166

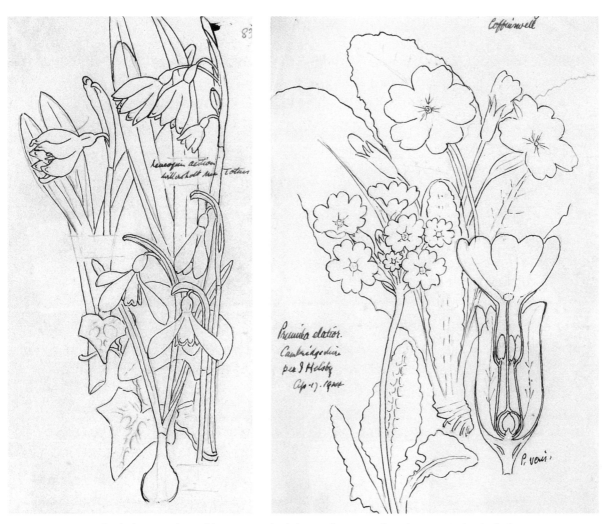

Pen and ink drawings by Keble Martin which he used to group his plants on individual plates.
LEFT Snowdrop (*Galanthus nivalis*). See plate 83 Iridaceae. RIGHT Primrose (*Primula vulgaris*).
See plate 57 Primulaceae. By permission of the Linnean Society of London.

easily appears a great deal of time has not been spent in composing it; indeed only a few leisure hours in the winter, for want of company, by way of diversion, not at all interfering with, much less interrupting my proper studies, or the necessary attendance on the duties of my parish, which I think ought always to be uppermost.

These are sentiments that Gilbert White would have warmly endorsed. He extended his garden at Selborne, introducing a ha-ha, walls, a hermitage, a zigzag path up the slope of The Hanger, and meticulously maintained a 'Garden-Kalendar'. Some clerics chose to specialize in particular families, genera or groups of plants. The Revd William Herbert (1778–1847) hybridized species of Amaryllidaceae. Also a competent artist, he drew more than 100 plates for *Curtis's Botanical Magazine* and the *Botanical Register*. A passionate rosarian, the Revd Samuel Reynolds Hole (1819–1904) tended more than 4000 rose bushes at Caunton, where he was curate and vicar for more than 40 years. His *Book about roses* (1869) went into many editions. Two books by the Revd Henry N. Ellacombe (1822–1916), *In a Gloucestershire garden* (1895) and *In my vicarage garden and elsewhere* (1902), affirm his devotion to a garden which kept him from moving to any other parish.

167

Such an attachment to a garden or a village or neighbourhood was not that unusual. The Revd F.O. Morris remained nearly 40 years at Nunburnholme; the Revd Octavius Pickard-Cambridge, who followed his father as Rector of Bloxworth, stayed there for almost half a century; but the prize for longevity of tenure must surely go to the Revd William Kirby (1759–1850), who ministered to the needs of his parishioners in Barham for 68 years.

What these clergymen and all the authors and artists discussed in this book had in common was an awareness and an appreciation of nature. They set out to write or to paint primarily to communicate information or pleasure to the reader. Of course, personal recognition and profit were additional incentives.

Tenacity in the face of difficulties was a quality they all shared. Blindness and other disasters never daunted Rumphius. Catesby learned how to etch when he could not afford an engraver. Mrs Blackwell was another paragon of self-reliance. Losing money on the *Flora Londinensis* did not deter Curtis from venturing upon another publication. Thornton beggared himself in his determination to make *The Temple of Flora* a work that upheld British superiority in book production.

Patronage helped a struggling author and subscribers were crucial for a book's success. Proposals announcing the advent of a new book were printed and distributed, but publicity was only the first stage in the enticement of subscribers, who would quickly defect if publishing schedules flagged or standards of production dropped. Relations between authors and booksellers/publishers were not always harmonious, the former convinced that the latter always benefited most from any contract. The production of books demanded dependable collaboration between author, artist, engraver, colourist and publisher. Engravings absorbed much of the production costs and lavishly illustrated books seldom paid their way. Print runs were small, inflating unit costs. The sale of fine and prestigious books suffered in times of war and economic recession. With all these obstacles we can only marvel that so many superb natural history books were created.

Postage stamps designed by Keble Martin: Hawthorn, Bindweed, Ox-eye or dog daisy, Bluebell.
Issued in April 1967. Philatelic Collection, B.L.

Bibliography

Adams, B. *The flowering of the Pacific: being an account of Joseph Banks' travels in the South Seas and the story of his florilegium*. London, Collins and Natural History Museum, 1986.

Addison, W. *The English country parson*. London, Dent, 1947.

Allen, D.E. *The naturalist in Britain: a social history*. London, Lane, 1976.

Allen, D.E. 'Natural history in Britain in the eighteenth century', *Archives of natural history*, vol. 20, part 3, 1993, pp.333–47.

Altick, R.D. *The English common reader: a social history of the mass reading public, 1800–1900*. University of Chicago Press, 1957.

Altick, R.D. *The shows of London*. Harvard University Press, 1978.

Anker, J. 'From the early history of the Flora Danica', *Libri*, vol. 1, 1951, pp.334–50.

Armstrong, P. *The English parson-naturalist: a companionship between science and religion*. Leominster, Gracewing, 2000.

Audubon, M.R. *Audubon and his journals*. London, Nimmo, 1898. 2 vols.

Aymouin, G. *The Besler florilegium: plants of the four seasons*. New York, Abrams, 1987.

Banks, Sir J. *The Endeavour journal of Joseph Banks, 1768–1771*. Edited by J.C. Beaglehole. Sydney, Public Library of New South Wales, 1962. 2 vols.

Banks, Sir J. *Captain Cook's florilegium: a selection of engravings from the drawings of plants collected by Sir Joseph Banks and Daniel Solander on Captain Cook's first voyage to the islands of the Pacific, with accounts of the voyage by Wilfrid Blunt and of the botanical explorations and prints by William T. Stearn*. London, Lion and Unicorn Press, 1973.

Barber, G. 'Books from the old world and for the new: the British international trade in books in the eighteenth century', *Studies in Voltaire and the eighteenth century*, vol. 151, 1976, pp.185–224.

Barber, G. 'Book imports and exports in the eighteenth century', in R. Myers and M. Harris (editors), *Sale and distribution of books from 1700*. Oxford Polytechnic Press, 1983.

Barker, N. *Hortus Eystettensis: the bishop's garden and Besler's magnificent book*. London, The British Library, 1994.

Barnes, J.J. *Free trade in books: a study of the London book trade since 1800*. Oxford University Press, 1964.

Beekman, E.M. (editor) *The Amboinese curiosity cabinet. Translated, edited and annotated with an introduction by E.M. Beekman*. Yale University Press, 1999.

Blangrund, A. and Stebbins, T.E. (editors) *John James Audubon: the watercolours for The Birds of America*. New York, Villard Books, 1993.

Bleiber, E.F. (editor) *Early floral engravings: all 110 plates from the 1612 'Florilegium'*. New York, Dover Publications, 1976.

Blum, A.S. *Nineteenth-century zoological illustration*. Princeton University Press, 1993.

Bradbury, H.R. *Nature printing: its origin and objects. A lecture as delivered at the Royal Institution of Great Britain … on … May 11, 1856*. London, Bradbury and Evans, 1856.

Bridson, G.D.R., Wendel, D.E. and White, J.J. *Printmaking in the service of botany*. Pittsburgh, Hunt Library for Botanical Documentation, 1986.

British Museum (Natural History) 'Catalogue of the natural history drawings commissioned by Joseph Banks on the Endeavour voyage 1768–1771, held in the British Museum (Natural History)'. *Bulletin, British Museum (Natural History), historical series*, vol. 11, 1984; vol. 12, 1987.

British Museum (Natural History) *The florilegium: Cook, Banks, Parkinson 1768–1771. Introduced by Hank Ebes*. London, British Museum (Natural History), 1988.

Britten, J. *Illustrations of Australian plants collected in 1770 during Captain Cook's voyage round the world in H.M.S. Endeavour*. London, British Museum, 1900–05. 2 vols.

Bruce, M.R. 'John Sibthorp'. *Taxon*, vol. 19, 1970, pp.353–62.

Bush, C. 'Erasmus Darwin, Robert John Thornton and Linnaeus' sexual system', *Eighteenth century studies*, vol. 7, 1973–74, pp.295–320.

Caldwell, J. *Sketches for the flora [of] W. Keble Martin*. London, Joseph, 1972.

Cardew, F.M.G. 'Dr Thornton and the "New Illustrations" 1799–1807', *Journal of Royal Horticultural Society*, vol. 72, 1947, pp.281–85, 450–53.

Carr, D.J. (editor) *Sydney Parkinson: artist of Cook's Endeavour voyage.* London, British Museum (Natural History), 1983.

Carter, H. *Sir Joseph Banks 1743–1820.* London, British Museum (Natural History), 1988.

Cave, R. and Wakeman, G. *Typographia naturalis.* Wymondham, Brewhouse Press, 1967.

Chalmers, J. 'Audubon in Edinburgh'. *Archives of natural history*, vol. 20 part 2, 1993, pp.157–66.

Colloms, B. *Victorian country parson.* London, Constable, 1977.

Corning, H. (editor) *Letters of John James Audubon, 1826–1840.* Boston, Club of Odd Volumes, 1930. 2 vols.

Curtis, S. *General indexes to the plants contained in … the Botanical Magazine to which are added a few interesting memoirs of the author, Mr W. Curtis.* London, Curtis's Botanical Magazine, 1828.

Curtis, W.H. *William Curtis 1746–1799: botanist and entomologist.* Winchester, Warren, 1941.

Curtis, W.H. 'George Graves (1784–1839)', *Watsonia*, vol. 2, 1951–53, pp.93–99.

Curwen, H. *A history of booksellers: the old and the new.* London, Chatto and Windus, 1843.

Darlington, W. *Memorials of John Bartram and Humphry Marshall with notices of their botanical contemporaries.* Philadelphia, Lindsay and Blakiston, 1849.

Desmond, R. 'William Roxburgh's Plants of the coast of Coromandel', *Hortus aliquando* no. 2, 1977, pp.23–41.

Desmond R. *A celebration of flowers: two hundred years of Curtis's Botanical Magazine.* London, Collingridge, 1987.

Desmond, R. *The European discovery of the Indian flora.* Oxford University Press, 1992.

Desmond, R. *Dictionary of British and Irish botanists and horticulturists.* London, Taylor and Francis, 1994.

Dunthorne, G. *Flower and fruit prints of the eighteenth and early nineteenth centuries.* London, Dulau, 1938.

Edmonds, J.M. and Powell, H.P. 'Beringer "Lügensteine" at Oxford', *Proceedings of Geological Association*, vol. 85, 1974, pp.549–54.

Feather, J. 'The commerce of letters: the study of the eighteenth-century book trade', *Eighteenth century studies*, vol. 17, 1983–84, pp.405–24.

Fontaines, U. de 'The Darwin service and the first printed floral patterns at Etruria', *Proceedings of the Wedgwood Society*, no. 6, 1966, pp.69–90.

Ford, A. *Audubon: a biography.* New York, Abbeville Press, 1988.

Foshay, E.M. *John James Audubon.* New York, Abrams, 1997.

Foster, P.G.M. *Gilbert White and his records: a scientific biography.* London, Christopher Helme, 1988.

Fournier, M. 'Enterprise in botany: Van Reede and his Hortus Malabaricus', *Archives of natural history*, vol. 14, part 2, 1987, pp.123–58; vol. 14, part 3, 1987, pp.297–338.

Freeman, R.B. *British natural history books, 1495–1900.* Folkestone, Archon Books, 1980.

Frick, G.F. and Stearns, R.P. *Mark Catesby: the colonial Audubon.* University of Illinois, 1961.

Fries, W. *The double elephant folio: the story of Audubon's Birds of America.* Chicago, American Library Association, 1973.

Grandjean, B.L. *The Flora Graeca.* Copenhagen, Hassing, 1950.

Grigson, G. *Thornton's Temple of Flora … with bibliographical notes by Handasyde Buchanan.* London, Collins, 1951.

Harvey, J. *Early nurserymen.* London, Phillimore, 1974.

Hemsley, W.B. 'Robert John Thornton', *Gardener's chronicle*, vol. 2, 1894, pp.89–90.

Hemsley, W.B. *A new and complete index to the Botanical Magazine … to which is prefixed a history of the magazine.* London, Lovell Reeve, 1906.

Heniger, J. *Hendrik Adriaan van Reede tot Drakenstein (1636–1691) and Hortus Malabaricus: a contribution to the history of Dutch colonial history.* Rotterdam, Balkema, 1986.

Henrey, B. *British botanical and horticultural literature before 1800.* Oxford University Press, 1975. 3 vols.

Henrey, B. *No ordinary gardener: Thomas Knowlton, 1691–1781.* Edited by A.O. Chater. London, Natural History Museum, 1986.

Hunt Botanical Library. *Catalogue of botanical books in the collection of Rachel McMasters Miller Hunt. Compiled by Jane Quinby and Allan Stevenson.* Pittsburgh, 1958–61. 3 vols.

Hyde, R. 'Robert Havell Junior, artist and aquatinter', in R. Myers and M. Harris (editors) *Aspects of the English book trade*, 1984, pp.80–108.

Jackson, B.D. 'Dr Alexander Blackwell (1700–1747)', *Journal of botany*, vol. 48, 1910, pp.193–95.

Jackson, C.E. 'The changing relationship between J.J. Audubon and his friends P.J. Selby, Sir William Jardine and W.H. Lizars', *Archives of natural history*, vol. 18, part 3, 1991, pp.289–307.

Jahn, M.E. and Woolf, D.J. (editors) *The lying stones of Dr Johann Bartholomew Adam Beringer being his Lithographiae Wirceburgensis.* University of California Press, 1963.

Jahn, M.E. 'Dr Beringer and the Würzburg "Lügensteine"', *Journal of Society for the Bibliography of Natural History*, vol. 4 part 2, 1963, pp.138–46.

King, R. *The Temple of Flora.* London, Weidenfeld and Nicolson, 1981.

Lack, H.W. and Baer, W. 'Pflanzen aus Kew auf Porzellan aus Berlin', *Willdenowia*, vol. 6, 1986, pp.285–312.

Lack, H.W. and Ibanez, V. 'Recording colour in late eighteenth-century botanical drawings: Sydney Parkinson, Ferdinand Bauer and Thaddäus Haenke', *Curtis's Botanical Magazine*, vol. 14 part 2, 1997, pp.87–100.

Lack, H.W. 'Recording form in early nineteenth-century botanical drawings. Ferdinand Bauer's "cameras"', *Curtis's Botanical Magazine*, vol. 15, part 4, 1998, pp.254–74.

Lack, H.W. and Mabberley, D.J. *The Flora Graeca story: Sibthorp, Bauer and Hawkins in the Levant.* Oxford University Press, 1999.

Lambourne, M. 'John Gould and Curtis's Botanical Magazine', *Kew magazine*, vol. 11, part 4, 1994, pp.186–97.

Ledger, A.P. 'Derby "Botanical" dessert services, 1791–1811', *Derby Porcelain International Society journal*, vol. 2, 1991, pp.79–102.

Lee, H. *The vegetable lamb of Tartary: a curious fable of the cotton plant.* London, Sampson Low, etc., 1887.

Littger, K. and Dressendörfer, W. *The garden at Eichstätt: the book of the plants by Basilius Besler.* Cologne, Taschen, n.d.

Mabberley, D. *Ferdinand Bauer: the nature of discovery.* London, Holberton, 1999.

Mabey, R. *Gilbert White.* London, Century, 1986.

McBurney, H. *Mark Catesby's natural history of America. The watercolours from the Royal Library, Windsor Castle.* London, Holberton, 1997.

Mallatt, J.M. 'Dr Beringer's fossils: a study in the evolution of scientific world view', *Annals of science*, vol. 39, 1982, pp.371–80.

Manilal, K.S. (editor) *Botany and history of Hortus Malabaricus.* Rotterdam, Balkema, 1980.

Martin, W.K. *Over the hills.* London, Joseph, 1968.

Merrill, E.D. *An interpretation of Rumphius's Herbarium Amboinense.* Manila, Bureau of Printing, 1917.

Munby, F.A. and Norrie, J. *Publishing and bookselling.* Edition 5. London, Cape, 1974.

Myers, R. and Harris, M. (editors) *Sale and distribution of books from 1700.* Oxford Polytechnic Press, 1983.

Myers, R. and Harris, M. (editors) *A genius for letters: booksellers and bookselling from the sixteenth to the twentieth century.* Winchester, St Paul's Bibliographies, 1995.

Nance, E.M. *Pottery and porcelain of Swansea and Nantgarw.* London, Batsford, 1942.

Noblett, W. 'Pennant and his publisher: Benjamin White, Thomas Pennant, and Of London', *Archives of natural history*, vol. 11, part 1, 1982, pp.61–68.

Noblett, W. 'Dru Drury, his Illustrations of natural history (1770–82) and the European market for printed books', *Quaterendo*, vol. 15, 1985, pp.83–102.

Perkins, W.F. 'Dr Thornton's works', *Gardener's chronicle*, vol. 2, 1894, pp.276–78.

Pieters, F.F.J.M. and Winthagen, D. 'Maria Sibylla Merian, naturalist and artist (1647–1717): a commemoration on the occasion of the 350th anniversary of her birth', *Archives of natural history*, vol. 26 part 1, 1999, pp.1–18.

Plant, M. *The English book trade: an economic history of the making and sale of books.* London, Allen and Unwin, 1974.

Ralph, R. *William MacGillivray: creatures of air, land and sea.* London, Holberton, 1999.

Raven, C.E. *John Ray, naturalist: his life and works.* Cambridge University Press, 1942.

Royal Botanic Garden, Edinburgh and Danish Cultural Institute, Edinburgh *Flora Danica.* Edinburgh, 1994.

Rucker, E. and Stearn, W.T. *Maria Sibylla Merian in Surinam. Commentary to the facsimile edition of Metamorphosis insectorum Surinamensium. Based on watercolours in the Royal Library, Windsor.* London, Pion, 1982.

Schaaf, L. 'The first photographically printed and illustrated book', *Papers of Bibliographical Society of America*, vol. 73, part 2, 1979, pp.209–24.

Schaaf, L. 'Anna Atkin's cyanotypes', *History of photography*, vol. 6, 1982, pp.151–72.

Schaaf, L. *Sun gardens: Victorian photograms by Anna Atkins.* New York, Aperture, 1985.

Sirks, M.J. 'Rumphius, the blind seer of Amboina', in P. Honig and F. Verdoon *Science and scientists in the Netherlands Indies*, New York, Board for Netherlands Indies, 1945, pp.295–308.

Sitwell, S., Buchanan, H. and Fisher, J. *Fine bird books, 1700–1900*. London, Collins, 1953.

Sitwell, S. and Blunt, W. *Great flower books, 1700–1900*. London, Collins, 1956.

Smith, Sir J.E. *A selection of the correspondence of Linnaeus and other naturalists*. London, Longman, 1821. 2 vols.

Smith, Lady P. *Memoir and correspondence of Sir James Edward Smith*. London, Longman, 1832. 2 vols.

Stearn, W.T. 'Sibthorp, Smith, the "Flora Graeca" and the "Florae Graecae prodromus"', *Taxon*, vol. 16, 1967, pp.168–78.

Stearn, W.T. 'From Theophrastus and Dioscorides to Sibthorp and Smith: the background and origin of the Flora Graeca', *Biological journal of the Linnean Society*, vol. 8, 1977, pp.285–98.

Stearn, W.T. 'Maria Sibylla Merian (1647–1717) as a botanical artist', *Taxon*, vol. 31, part 3, 1982, pp.529–34.

Stearn, W.T. *Plant portraits from the Flora Danica, 1761–1769*. Mendip Press, n.d., pp.4–7.

Synge-Hutchinson, P. 'Sir Hans Sloane's plants and other botanical subjects on Chelsea porcelain', *Connoisseur year book*, 1958, pp.18–25.

Tjaden, W.L. 'Herbarium Blackwellianum emendatum et auctum … Norimbergae, 1747–1773', *Taxon*, vol. 21 part 1, 1972, pp.147–52.

Tomasi, L.T. *An Oak Spring flora. Flower illustration from the fifteenth century to the present time. A selection of the rare books, manuscripts and works of art in the collection of Rachel Lambert Mellon*. Upperville, Virginia, Oak Spring Garden Library, 1997.

Turner, D. *Extracts from the literary and scientific correspondence of Richard Richardson*. Yarmouth, 1835.

Valliant, S. 'Maria Sibylla Merian: recovering an eighteenth-century legend', *Eighteenth century studies*, vol. 26, 1992–93, pp.467–79.

Wagner, P. and others *Flora Danica and the Royal Danish Court*. Copenhagen, Christiansburg Palace, 1990.

Wagner, P. 'Icones Florae Danicae. The "plant illustrators" of Flora Danica and the "school of illumination for women"', in V. Woldbye (editor) *Floral motifs in European painting and decorative arts*. The Hague, SDU Publishers, 1991.

Wakeman, G. 'Henry Bradbury's nature printed books', *The library*, vol. 21, 1966, pp.63–67.

Wallis, P.J. 'Book subscription lists', *The library*, vol. 29, 1974, pp.255–56.

Wettengl, K. (editor) *Maria Sibylla Merian: artist and naturalist*. Frankfurt, Verlag Gerd Hatje, 1998.

Wijnands, O. 'Portraying new plants in the Renaissance', in *Royal Horticultural Society proceedings of first symposium on botanical art*, 1993, pp.14–22.

Wiles, R.M. *Serial publication in England before 1750*. Cambridge University Press, 1957.

Wilson, D. 'The iconography of Mark Catesby'. *Eighteenth century studies* vol.4, 1970–71, pp.169–83.

Wit, H. C. D. de *Rumphius memorial volume*. Baarn, Vitgeverij en Drukkerij Hollandia, N.V., 1959.

Index

Page references to illustrations are in italics.